JEAN-PIERRE AND LUCILE GARNIER MALET

THE POWER OF THE INVISIBLE

Those Hidden Truths

The Science of Time Doubling

reconocerse

First book by the same authors:—
Change Your Future Through Time Openings

Second book:
The Double... How Does It Work?

Translation: Carolina Rosset Gómez Publisher: Carolina Rosset Gómez
https://editorialreconocerse.com

First edition: May 2026

ISBN: 978-84-942181-4-9

Deposit: T1005-2026

Printed by Romanyà/Valls, s.a. Verdaguer, 1 -
Capellades (Barcelona)

Contents

A dream that could come true

Changing our future is everyone's dream... but, is that possible?

In the universe, everything is doubled: particles, stars and galaxies – so why shouldn't we be too?

A particle falling into a black hole disappears instantly to an outside observer. But if this observer is inside the black hole, he would notice that it takes an infinitely long time for the particle to enter it. This time differentiation also applies to twins, where one of them leaves in a rocket. Due to his speed, he would not age as quickly as the other who remained on Earth. First proposed by Paul Langevin in 1922, this paradox was experimentally proven in 1972.

This twin could depart in zero time and return instantaneously, after having travelled for a long time in another time. Initially defined by Poincaré and Einstein, might this relativity of time not be at the service of mankind?

Let us therefore agree that these null times, or more precisely imperceptible times, exist. Medical imaging has already revealed to us that imperceptible subliminal images modify our memory. This modification would in fact be the result of long experiences carried out in another reality, by observers evolving in another time.

It is the physical principle of time doubling that we could follow, to allow us to survive and create our future possibilities and anticipate our future before living it. But how? Due to the existence of a double in another time.

Now, without knowing either of its existence or its purpose, we are living through the end of a time doubling cycle, governed by the planetary movements of our solar system, which is currently opening all the doors of information to us.

This universal movement is found equally in galaxies and in our DNA. Because of it, we benefit at every moment from two doubled times in which we live simultaneously.

To find, in this "end (of doubling) of times", a personal and planetary balance: Change your future through time openings, is a vital principle that would be good to know how to use.

This story opens a window to a reality that some prefer to keep firmly shut. It could disturb traditionalists of all kinds, who do not wish to give up their current mindsets, much less evolve them.

I thank my wife for helping me to reveal, in the form of a fable, essential truths forgotten in the night of time. Let us try to delve deeper into them so that we may cease to be troubled by useless postulates and dogmas, the bitter fruits of centuries of ignorance!

At last, the invisible is within our reach: let's use it in our daily lives instead of striving to ignore its power!

JEAN PIERRE GARNIER MALET

1

The man on the park bench and the beggar

Louisiana walks through the park near her home: she feels happy, as she is returning from her bank where the manager has granted her a loan to settle her overdraft. She thinks that her publisher, Marc, will soon give her an advance on her next book and everything will be back in order. She is jumping for joy inside. She can continue telling anyone who will listen: "There are no problems, only solutions!"

Her joy doesn't last long, as a child throws a ball that hits her right in the face, causing her to drop her mobile phone. As she picks it up from a flowerbed, without realising it, she sinks her left shoe into sticky mud. She retrieves it, cleans it as best she can on the grass, and puts it back on. The bored expression on the boy's face as he walks away, still looking at her, makes her smile despite everything. But why is it that joy is always followed by misfortune? She sits down on a bench to call Marc. She is worried about the project they are working on; she doesn't know what to tell him. The manuscript is far from finished.

Lost in her thoughts, she stares at her phone screen just as a stranger speaks to her, startling her. She hasn't seen or heard him approach, but he had sat down there, beside her, without making a sound. As she discovers this man in his thirties, she is astonished and bewildered when he starts a conversation:

"Why waste time on a mobile when you can talk to the birds? They can answer your questions much better than your friend Marc."

Talk to the birds! Is this stranger making fun of her? Louisiana's surprise, tinged with a dose of bewilderment, turns to suspicion. How could this stranger know she was going to call Marc? This surely, can't be a coincidence.

"I know," says the stranger, "because I'm the one suggesting your questions to you. I know your questions because they're related to mine."

That sharp tone does not reassure her. The shivers that were running down her spine now course through her entire body.

"Do we know each other?" asks Louisiana, bewildered.

"Look! Look at those birds! They know me well. They don't need to see me to know I'm here..."

Louisiana would like to stand up but cannot; as if she were strangely reassured by her fear, a paradoxical source of comfort and well-being.

The sparrows come to peck at the grass around them. Emboldened, one of them approaches whilst...inspects Louisiana's shoes; she leans forward, smiling, ready to stroke it. But the bird takes flight, wary, dragging the others after it.

The young man sighs, shaking his head. This exasperates Louisiana even more. A sparrow returns towards her, but suddenly turns away as if suddenly afraid.

"Your thoughts are all over the place and the birds can tell your actions don't match your thoughts. You frighten them! Stroke them when they ask you to!'

This young man's self-assurance intrigues Louisiana to her core.

“What I’m saying isn't funny at all.”

This way of guessing things continues to surprise Louisiana, who stares intently at this stranger, not daring to speak... or even to think...

"If it wants to be petted", says the stranger with an amused air, ‘the bird lands on the hand of the one who wants to pet it, asking nothing in return, just for the emotion of the moment, because emotion allows us to exchange our joys to ease our sorrows...’

The young man continues:

“Why do you think I must be part of a sect? Some very strange thoughts are running through your mind.”

Stunned by the fact that this man can read her thoughts so easily, Louisiana stares into his eyes, trying to understand this curious one-way telepathy.

"But who are you?" asks Louisiana, perplexed.

Without answering, he extends his hand in front of him and, instantly, a bird tries to land on it but seems to pass right through.

“I think you should put Marc aside and restart your manuscript from the beginning. You must first interest those who seek the purpose of life on earth and want to
understand the usefulness of your book in daily application.”

Looking concerned, he looks at Louisiana before continuing:

“You know well that in the scientific community, the ridicule of a new theory often only interests critical minds who try to delve into its truth and not into the Truth.”

Louisiana is left speechless. Her surprise paralyses her so much that she decides to leave. She stands up, lost in her thoughts; her steps becoming urgent, her questions multiplying. She wanders aimlessly until, exhausted, she collapses at the end of a bench without realising that a beggar is already sitting on it.

“Don’t you have a euro to spare for someone who accepts you beside him?”

This boldness plunges Louisiana back into her thoughts, especially when he asks her in a slurred voice, due to the excess of alcohol.

"You see, litttle mother, you should get it into your head that I am the man on the bench, now and always. The rent is ten for a quarter of an hour."

Astonished, Louisiana doesn’t react; she remains silent, disturbed by such self-assurance. A voice pulls her out of her disconcerting lethargy:

"Well, are you going to cough up the cash?"

Louisiana instinctively turns towards the speaker. She is greatly surprised. It is, in fact, an older, unshaven, dishevelled and much dirtier version… the almost exact replica of the ‘young man from the park bench a while ago’.

“It’s been three minutes now, sweetheart, you owe me two euros.”

At first bewildered by the resemblance, then amused by his audacity, and wanting to heighten the effect of the unexpected, Louisiana rummages through her purse.

“I’ve only got fifty euros.”

“Don’t worry,” replies the beggar, snatching the note from her hands. “I’ve got change... I might not know how to save, but I know how to anticipate!”

And with a hand that probably hasn’t seen water for ages, he pulls out a bag from his pocket and empties it onto his knees. It’s his change... all his change, to be more precise... that is, three or four euros... Louisiana refuses and demands her note back.

“For you,” says the beggar, “this is superfluous. For me, it’s vital!”

And without a shred of embarrassment, the homeless man puts the change and the note back in his pocket.

"They’re yellow coins," he explains, flashing a hideous set of teeth as he laughs, "nothing worth mentioning compared to your big note."

He bursts out laughing.

Louisiana doesn’t like this situation at all.

‘My note is called “return”’ she begins timidly.

“All the passers-by who share their immense generosity of spirit with me always add a speech for the occasion... It’s the charity of bastards! Don't tell me you have nothing to say to compensate for your generosity..."

Without waiting for Louisiana's reply, he adds:

"Above all, don't talk to me about global warming, like all those snobs who've never sat their backsides on a bench at night."

At that very moment, a passer-by drops a five-euro note into the beggar's greasy hat.

He picks it up and says to Louisiana:

"There's another bastard easing his conscience..."

Noticing the questioning look on Louisiana's face—who, in reality, is only worried about her fifty euros—he continues:

"I bet he just cheated on his wife with a hundred-euro whore. Or maybe it was his best friend's wife, or maybe..."

Seeing that she looks even more surprised, he adds:

"You too think that a beggar like me has no past. People tend to think that one is born a beggar!"

He hands her back her fifty-euro note.

"Take your note! You want to teach me how to live when you are the one who should be learning from me! Do you know why I'm the man on the bench now and forever?"

Louisiana makes no attempt to stop the beggar's verbal diarrhea. She listens attentively out of politeness, glad to have recovered her note.

“Right here where you see me, I used to be a history teacher, ancient and modern history, my dear, yes sir!”

Louisiana nods, perplexed. The beggar persists and declares:

"Shaven, perfumed, in a suit and tie, with polished shoes, I was a tenured professor at the Sorbonne. I was devastated by the death of my wife, run over by a bastard. We were inseparable... what despair! A nail that felt bad when separated from its flesh. Our resemblance lay only within us. Do you know what death is? A bitch, a whore who took my other half. We felt strong and invincible... We were two fools, happy to be together. She had some incredible ideas about life, you can’t even imagine. But life didn’t want to think about her. A bitch, too! That’s all! See, you’re making me cry..."

Tears roll down his dirty cheeks, leaving greasy streaks. He doesn’t even try to wipe them away, as his thoughts are wandering elsewhere...

Louisiana doesn’t know what to say. She’d like to leave, but thinks it wouldn’t be a good idea.

“She was the one who had the money, the flat, everything... all I had left was my salary and the bottle... but it wasn’t History with a capital H, the one that could give me
back the happiness I’d lost in my own little shitty story.”

The homeless man’s filthy appearance gave no hint whatsoever of a glorious past. It was hard to imagine a history teacher reduced to such a wretched state.

The homeless man wiped his tears with the back of his sleeve, leaving even more smudges on his face.

"Then, after countless failed projects, I ended up on this bloody bench watching other bastards go by. But I'm happy here, sweetheart, because I don't depend on anyone anymore. I speak as I please, rudely or respectfully depending on the circumstances. And like Diogenes, I can say to the cops who look at me suspiciously: "Gentlemen, get out of my way!" Come on, go ahead with your moralizing speech, I'm listening! "Why do you think I'm going to give you a moralizing speech?"

"If you sit on my bench, it's because you're not normal! And on top of that, I'm starting to like you, especially if you give me your fifty euros again! In which case, I'll tell you how to scare your banker.

This remark hits Louisiana like a ton of bricks; she is genuinely worried about the bank manager's comments.

"And in return?" replies Louisiana, amused.

"Listen to this story which, in this world of bankers and swindlers, is worth more than your fifty euros. It's a fable worthy of Aesop, set by yours truly. A shrewd man decides to buy the donkeys' oats at a hundred euros a kilo, an exorbitant and utterly crazy price.The donkeys take advantage and sell it despite it beeing their favourite food. He promises to return the following week to buy more, this time at three hundred euros a kilo. But, unfortunately, the donkeys had no more oats left to sell. The cunning man therefore sends them an offer to sell them, at two hundred and fifty euros a kilo, the oats he has just bought from them for a hundred. Naturally, the donkeys fight to buy it. The cunning man thus makes a profit of one hundred and fifty euros per kilo. And this trade continues until one day, with no oats left, the donkeys starve to death because of their greed."

Returning to his coarse tone, he adds:

"You can see my story's well worth your fifty euros. So get off my bench if you don't want to pay the rent!"

Louisiana feels a bit silly… but, glad to have gotten off lightly, she quickly leaves. She just wonders how she could have sat down so absent-mindedly on that bench, without noticing the presence of the beggar, who was quite annoying, by the way.

Well, never mind! Nothing surprises her anymore.

“You've got a superiority complex… yet these are the men who can teach you something important.”

Louisiana turns abruptly at the sound of a familiar voice and spots the young man from the park bench from earlier. The resemblance to the homeless man seems even more surprising to her. Had he followed her here?

“No, I'm not following you, don't worry!”

Seeing the puzzlement on Louisiana's face, he quickly adds:

"You know, telepathy is much faster than walking, and if I look like the other guy, it's because you think so. Do you want to prove that you experience what you think? Then you should go back to where you came from to recover your fifty euros back.”

Louisiana glances at the beggar. Her surprise is immense. There is no one sitting on the bench anymore, but on the ground there is a fifty-euro note and two policemen looking at it. She hurries over to pick it up. The officers walk away. Just like the homeless man, the young man from the park bench has vanished.

She wonders if she's going mad... Is she having a waking nightmare? How on earth did her bill end up there? She runs towards the officers and asks them if they saw a man on the bench. They look at her, surprised by the question. Louisiana stammers:

"I gave my fifty euros to a homeless man who was sitting on the bench."

Seeing the officers' surprise, she clarifies:

"He wasn't begging, I was paying him for a place on the bench. He gave me my change and..."

The officers don't understand a word.

"It was a bet… I lost… I owed him that money," stammers Louisiana.

They realized Louisiana had been robbed, looked at her with a condescending air and walked away laughing.

As for her, she doesn't know what to think. Is she living in another reality?

Yes, but how could she imagine that a stranger might tell her about a new publisher she doesn't yet know, and at the same time intrude into her life?

No doubt, it's possible to live in the fairy tale we imagine; she knows that. But had she already imagined this story? Maybe... thinking about it, hadn't she wished, in her wildest dreams, that a publisher would come looking for her?

2

Dream or reality?

One by one, Louisiana goes over the events of the previous day in her head, very strange, not to say disturbing. She recalls every detail of the unsettling conversations with that unusual young man.

Had she been dreaming, or had she experienced a brief moment of unconsciousness? Now she regretted not having been up to the challenge. She believes, and has always told anyone willing to listen, that coincidence does not exist! How could she have let herself be taken by surprise in that way, and failed to seize the reins of a unique opportunity to reflect on those strange coincidences?

What is clear is that she had failed to make the most of that fortuitous and simply extraordinary encounter.

So, the next day, as she feels the fifty euros in her pocket, she decides to return to the famous bench.

***

The homeless man is standing there. Having decided to give him the ten euros for rent, Louisiana crosses the street. To her astonishment, the beggar hands back her purse, which he had surreptitiously stolen whilst listening to his story about the donkeys.

"There were ten euros... I drank them with delight, sensuality and love... I'm telling you this to show you that I'm not just some bastard sitting on my bench."

"I didn't say that, I just came back to pay you the rent, as I found the fifty euros under the bench." And she shows him the bill.

"I looked for it, you know?... I was furious. I thought the cops had picked up what I'd stolen from you dishonestly."

"Fifteen minutes rent is the ten euros you've already taken; we're even!"

The beggar, impassive, looks Louisiana up and down.

"Well, now that you think you've paid me, I have to listen to your nonsense, don't I?"

"Right, right... I'll leave you in peace!"

"What's going on? Are we not using the familiar form anymore?"

"Is that the rule between the thief and his victim?"

"The thing is, you've came back to look for your wallet..."

No, I thought I'd left it at home. Actually, I wanted to know if my thief is really real, so I'm leaving reassured."

The homeless man doesn't understand anything.

"If I am real?... But what on earth are you talking about?"

Don't worry, it's just a story between me and myself.

“You’re so out of it,” he says mockingly, “that I don’t really care to listen to your stories, you know?... This is different from the not-so-Catholic, or too popular help from the rich who think they’re the poor people’s little brothers... There’s so much bullshit that it’s easy to get lost.”

“Well, I’ve got nothing to say to you, apart from the fact that I think it’s truly stupid that a so-called ‘history teacher’ should be reduced to begging on a bench.”

"If you’ve got something that would allow me to live better, I’m all ears, my dear… go ahead with your ten-euro morality! and tell me how a ‘so-called’ history professor can change his miserable life with a wave of a magic wand!"

"Start by imagining a better place than this rotten bench, and you’ll live better!"

The homeless man doesn’t seem to be listening to her, as his pleased gaze ‘scans’ a beautiful young lady.

"And you want me to control my thoughts, which ejaculate prematurely in my head when I see such a beauty pass by!"

“Would you have preferred that I give you some money?"

"Yes, but in doing so, you would collapse all my
“aphro-di-si-ac” dreams."

Seeing that Louisiana isn’t reacting to his tasteless joke, he changes the subject:

"You’re using the familiar terms with me anymore?… because with that schoolmistress look on your face, I feel like I’m stuck to a school desk. But tell me, Miss Know-it-all, if I imagine that young woman smiling at me, could it happen?"

"Why not? Try it"

He bursts out laughing.
"So, I think I'm a bastard and I become a bastard. Is that your story?"

"No, when you already are one! Well, good luck!"

The homeless man's face no longer wears its mocking expression.

“You know, sweetheart,” he says, trying to keep her there, “I'd really like it if the bastards who pass in front of me would smile at me, instead of turning away their cod-fish eyes. If you can manage that, I'll do anything you want, promise! And if I'm lying, I'll go to the priests' paradise where, after a chaste and pure life, you eternally screw everything that stands on its hind legs."

Filled, however, with a certain compassion, Louisiana imagines him as a child, fascinated by anything. So she continues, reassuringly.

Just imagine if passers-by could read what you're thinking."

"They wouldn't have made the trip for nothing, those bastards!

"Yes, but if they'd been happy with the trip, as you say, they'd know that what you think of them suits them."

"OK, and then?"

"Then there are two possibilities," replies Louisiana as she sits down. "Normally, when you see a beggar sitting on a bench, you cross the street or put on a worried look as if you haven't seen anything, but you'd never sit down next to him to start a conversation."

Laughing at the looks from others, she continues talking, unaware that using the informal 'you' comes naturally to her.

"If the passer-by's thoughts match yours, you attract them like a magnet. They find you likeable and they want to please you. If you think a smile would please you, they smile at you but won't give you anything!"

"Why not?"

"Because you haven't thought that you'd like to receive a coin."

"But that's nonsense!"

"Yes, but that's how it is! And if you think they're mocking you with their smile, in that case, your thoughts become the opposite of theirs. You find them unpleasant, they sense it, and out of fear, they might give you a euro."

"That's true! I've noticed that some of them rummage in their pockets or make a show of it when I look them straight in the eye, trying to intimidate them or make them feel sorry. And sometimes they give me a coin, although they usually act as if they don't have a penny, with obvious bad faith."

"That's normal, because your thoughts aren't benevolent."

"Yes, but they're bene-voliant, because it's me who's starving, not them."

"You're starving because you want to."

The homeless man doesn't accept that remark and becomes aggressive.

"Are you laughing in my face?"

"Shall we try it?"

"Try what? Laughing in my face?"

"Watch your thoughts… Just to show you that if you've got nothing, it's your own fault!"

"But what are you talking about! I've tried everything…
You're not going to change the world with your stories."

Saying this, he puts his hat on the ground and adds four coins.

"Why are you doing that?"

"So that the bastards who give nothing can see there are fewer bastards."

Seeing that Louisiana sighs and gets ready to stand up, he softens.

"Yes, I know, I'm creating my future!"

"Then take off that disgusting cap; it puts passers-by off, just as your filth keeps me away from you."

"And do you think your thoughts are correct when you say this to me?"

"What would be wrong is to think it without telling you.

The beggar has an old newspaper. Louisiana tears off a page, folds it into a little bowl and places it on the ground in front of her.

"And do you think that if people see you dressed like a princess, they'll give you money?"

"Yes, but on one condition: that you keep your distance and watch your thoughts."

"But I can keep an eye on them from the bench…so, shall we?

"Let's go!"

He closes his eyes and chants:

"Everyone, how good they are! Everyone, how lovely they are! Everyone, how good they are! Everyone…"

"Saying it is no use. You have to think it!… Look at that couple coming over."

Suddenly, the man stops, searches in his pockets whilst talking to his wife, who opens her handbag. They talk as they approach slowly. When they reach Louisiana, the woman hands her partner a five-euro note. He bends down to place it gently inside the little bowl. The couple walks away without waiting for so much as a thank you. They would have had to wait a long time for it, as the beggar is left open-mouthed, but he concentrates again.

Right across from the bench, a woman comes out of her shop and heads straight toward Louisiana:

"Can't you go and talk somewhere else? You're scaring my customers away."

Probably used to such reprimands, the homeless man isn't intimidated:

"This is my home, mate, so go back to your shop, clean your shoes and leave us in peace!"

The shopkeeper explodes:

"I'm going to call the police and you'll see!"

She returns to her shop, full of anger.

"And do you think that's being benevolent, talking like that?"

Louisiana asks, surprised.

"Of course it is! She won't stop nagging us; we have to set boundaries, everyone in their own home, for goodness' sake! Here come three benevolent people!"

Three friends approach, completely absorbed in a discussion. One of them takes out a purse and leaves 2 euros on the bill while smiling at the homeless man. Continuing their conversation, the other two look at him with a sigh, thus expressing their disapproval.

The homeless man picks up the bill. Louisiana speaks up.

'Why take it away?"

"Someone might pick it up as they walk past."

"You haven't understood a thing. You're creating the thief and the urge to steal from you."

"Haven't you seen the other two? Boy, were they keen to pick up their friend's money!"

"Without him, they wouldn't even have noticed us. If you think they're going to steal from you, your thoughts attract the thieves who will make use of the future you've created."

Louisiana takes the bill from his hands, bends down to put it back under the coin in the paper pot, which she smooths out to make it look better.

"The harvest will be much better!" she says, straightening up.

"What are you playing at?"

The two officers are standing there in front of her, looking very suspicious.

"I was just putting this gentleman's banknote back in its place."

Louisiana is utterly taken aback. She is alone in the bank and senses that the officers are beginning to ask themselves questions.

"It's not mine! It belongs to the beggar who was standing next to me a moment ago. I was looking after his money while he..."

"Your ID, please!" demands the taller one, suspicious.

Louisiana knows full well that in the situation she finds herself in, it is very difficult to be convincing. She searches for her purse and discovers... that it has disappeared along with the fifty-euro bill.

3

The police Officer and an old friend

The police take her to the station while asking her the usual questions. Louisiana's answers leave them completely baffled, a feeling that is only exacerbated when the chief inspector himself asks to see her. Bewildered,
Louisiana sees Jacques sitting behind his desk.

"Jacques, what are you doing here?"

"I'm a police commissioner," he says with a tone of false seriousness.

Louisiana is stunned:

"You, a commissioner?"

"We need one! The commish serves… silly girls like you."

When they shake hands, this silly remark makes them recall many bad memories.

"Never, ever," exclaims Jacques with a hypocritical smile, "not even in my wildest dreams would I ever have imagined seeing such a troublemaker as you here!"

"Dear Jacadi, I return the compliment!" replies Louisiana, still reeling from the shock of the surprise.

“And to what do I owe the displeasure of your visit?”

"A forced visit, let’s not forget!"

"Forced, but probably not pointless, since this gives us the chance to detest each other even more, let’s not forget. What can I do for you?"

Jacques explains:

“Well, yes! I’m familiar with your father’s theory… I’ve even read his latest book… a surprising piece of literature."

Jacques was drawn to the theory because he knew Louisiana. But its application in everyday life hadn’t convinced him.

He recalls the lecture by that scientist, the author of the time doubling theory, which he attended. Sitting in the front row, next to Louisiana, he glared and asked a question with the malicious intent of interrupting the lecture.

"Let’s assume that our thoughts create a potential future in another time, and that, every night, during paradoxical sleep, I arrange that future before living it. But, why talk about a double?"

"Because only he can fix our dreams, and that requires a good night’s sleep so that we can benefit from this vital help."

"How can a respectable scientist mix science and religion? Leave the priests’ examinations of conscience to the priests!"

Furious, Jacques stands up and leaves.

Stunned, Louisiana had been left speechless, just as she is now, at the police station, where Jacques reappears, now trying to soften his criticisms:

"Your father has come up with an extraordinary scientific theory, so, since you're here, explain a bit more to me… you never know!"

But without waiting for a reply, he continues, self-assured:

"In one of his videos, he claims that our thoughts are energies that allow us to create our future instantly."

"That's correct."

“Has this energy been measured? No, never!”

The tone of the questioning is that of the tense American interrogations that Jacques loves. Louisiana sighs but remains calm.

“Can you suggest something practical I can do in the field," says Jacadi, annoyed, "on the street, anywhere. So you can see for yourself that your father is talking nonsense when he talks about the energy of the future?"

"What sort of practical application?' replies Louisiana, surprised by such a suggestion, "with police officers?"

“Are you joking? You'll screw up and leave me high and dry. Don't worry, I've got what it takes to show you that you're completely mistaken about these ideas from start to finish. We can even give it a go right now, to prove to you that your stories of clairvoyance and intuition are a world away from the way I see things—that way we won't waste any time!"

He opens a folder and takes out a photo.

“What can you tell me about this little girl?”

He places the photo in front of Louisiana. It is a lovely little blonde girl of two or three years old who has just gone missing. Given her parents’ wealthy circumstances, kidnapping is suspected. Louisiana stands her ground.

“My name isn’t Madame Irma, you so-called psychic!”

"Well, so what? You can see that your psychic stories are useless."

“If I imagine that girl is going to be killed, I create a dangerous future for her that, without me, might not even exist.”

“So, in your view, the best thing is to stay there doing nothing, with your butt glued to your seat?” replies Jacques.

"Not at all! If they ask for my help, I can help this girl and perhaps guess where she is, if guessing can truly help her."

"It’s obvious!

"For me, it’s only obvious if it seems obvious to my double."

“Your double? Only someone like you would talk about a double.”

Realizing the futility of such a discussion, Louisiana gets to the point.

"I thought you understood what I was talking about."

"Yes! You go on and on about how only our double can guide us towards a better future. For me, the future I’m worried about is that girl's. So, what exactly are you doing in this specific case to help me with your double?"

"If he thinks it's a good idea, he'll tell me where the girl is."

Jacques sighs.

"And what am I supposed to do in an emergency? Wait until she's doubly dead?"

"In an emergency, keep a close watch on your thoughts so that they don't make the situation worse."

Irritated, Jacques rolls his eyes.

"And do you think that will solve the problem of kidnappings?"

"Yes, if we wish to obtain helpful intuitions."

Jacques doesn't think at all like Louisiana and lets her know it "Right, well, go on then with your intuitions! What do they tell you in this specific case?"

Louisiana protests vigorously:

"I'm no guru! Find the solution you're asking me for yourself. Stop being so aggressive! Think about helping that girl, focus your thoughts on kindness, and you'll see how you receive the intuitive answer to your problem!"

"The intuitive answer," sighs Jacques, trying to control the anger that is taking hold of him. With their noses in the muck, the dogs have scoured the whole countryside, in vain… and you, with that little mouth of yours, come and tell me that I'm not thinking of helping that girl!"

"I'm not talking about the dogs; I'm talking about you."

"I've got my snout in the muck too, so let's drop this conversation, which hasn't brought me anything new so far."

4

Untimely Acknowledgments

The next day, happy as a child, Jacques calls Louisiana. He is not the same man anymore. Often, fools become intelligent when they realize that intelligent people are more foolish than they are.

"We found the girl… I want to thank you…"

“Thank me for what?”

“Come and see me when you can, I’ll explain! This thing of yours is brilliant!”

In apparent haste, Jacques hangs up the phone.

***

Having heard the good news on the radio, Louisiana hurries to the police station. And there, flanked by two guards, appears the beggar, whose eyes light up when he sees her.

"Hello, beautiful! You've come at just the right time, because these two won't believe me... They say I've stolen your purse and the fifty euros you gave me!"

"But I already told them that nothing was stolen from me!" she exclaims, turning towards the officers.

The two officers look at each other uneasily, since they were the ones who brought Louisiana to the station the day before because of those dam fifty euros. In their beaten-dog looks, she thinks she can guess that they don't believe they are being pursued a blind fate, but by 'bloody bad luck'. What on earth will the boss say now!

But she is completely mistaken, as the officers do not see the beggar. For them, he doesn't exist!

"Is it true that you gave your wallet to a beggar with the fifty euros to a homeless man?" one of them asks her, stunned by the gesture.

Louisiana doesn't have time to reply.

"You're not going to start with your stupid questions again!" says the other. "Once and no more!"

"You see," says the beggar to Louisiana, "money and bullshit don't make those who have too much happy. And you, why are you here? Are you now giving your sermons and preaching at the police station?"

"I'm just waiting for the commissioner."

The officers look at one another, worried.

Seizing a moment of hesitation, the beggar leads Louisiana towards the exit.

“Come on, hurry up, you should never stay in a henhouse when you’re not a hen… they don’t have the same fleas as we do!”

Fearing an immediate reaction from the police, the homeless man rushes towards the exit.

But the policemen just smile. Unaware that she is the only one who can see the homeless man, Louisiana feels terrible:

"Tell the commissioner that…"

"What do you want to tell me? " says Jacques as he enters.

“Because of your rather unusual thanks, I’d thought of coming to see you, but…"

“No buts! Come with me to the adjoining room,” he says, winking at the two guards, who are pleased and relieved to see their boss happy.

***

The adjoining room… as you’ve guessed, is the bar across the street. Any miscreant who, without realizing it, came in for a drink would quickly feel out of place. As she follows Jacques, Louisiana thinks she hears the beggar’s voice:

“What’s going on! Don’t you buy your friends around?”

Louisiana turns around, smiling, and sees that Jacques is pretending not to hear.
"Does it really bother you that I’m talking to a beggar?"

“No, why would it?”

But she knows full well that the boss's arrival at the bar, along with this beggar, is causing a stir. As soon as they sit down, a waiter approaches them.

“The usual, Commissioner?”

“Me too!” says the homeless man. “I’ll have the usual.”

He turns to Louisiana, as if nothing were amiss:

"And you too, lovely, ‘the usual’?" Without waiting for a reply, the homeless man concludes:

“Right then, three ‘usuals’.”

“What is ‘the usual’?” asks Louisiana.

“It’s a pastis!” replies Jacques, who, intrigued, immediately adds, “I didn’t know you talked to the homeless.”

“I’d prefer a coffee!” she says, changing the subject, “and above all, I’d like to know why you’re thanking me.”

"As surprising as they are unexpected,” replies the beggar.

"Indeed, just as surprising as they are unexpected!" repeats Louisiana with a smile.

“But it’s thanks to you that I found the girl!”

The coffee arrives, along with a “usual”. The homeless man grimaces, and Louisiana watches in utter amazement as he grabs Jacques’s glass and takes a huge swig, putting it back in its place straight away.

“There’s so much ice in this that the breathalyzer tests won’t even register! Can you get me another one, drier this time?”

“Another one, please,” she says to the waiter, “but drier!”

“Would you like a “ristretto” coffee?” asks the waiter, as he doesn't see the homeless man either.

“No, a pastis.”

“A pastis?” asks Jacques, surprised.

“Two extra measures of pastis and two less of water,” the homeless man specifies.

“A pastis that’ll make the breathalyzer tests go haywire!” she adds amused.

Jacques is stunned.

“Well,” asks Louisiana, “tell us, what are these surprising thanks of yours?”

Jacques finally explains that the little girl had gotten locked in an old house, abandoned for some time, at the back of the garden. The old door lock had jammed. Although she screamed and cried until she was hoarse, no one had heard her.

"The strangest thing about all this," the inspector admits humbly, "is that you’ve helped me with your intuitive topics."

The homeless man can’t believe his ears. As for Jacques, he even feels embarrassed by this confession.

"It was just as I was arriving at the scene that I remembered your advice: “Stop your aggressive thoughts! Think about helping this girl and you'll see that you'll get the intuitive answer to your problem!'"

The homeless man is completely stunned.

"A cop without aggressive thoughts, does that even exist?"

Jacques ignores him and turns to Louisiana:

"And the answer was: 'Nose in the shit, the dogs searched the whole countryside!' And no dog can find anything with its snout in manure, and the old house was the former pigsty of the farm... Do you follow me?"

The homeless man seizes the moment:

"Well, to celebrate, I'll have the "usual", which we've ordered but hasn't arrived yet."

He signals to the waiter as no one is listening to him.

"As soon as you left, I rang the girl's father. And that's it... thanks again and a thousand apologies."

"But I've got nothing to do with it."

The waiter arrives with the pastis and a jug of water, which he places in front of Louisiana.

"You had the right intuition, that's all!"

The commissioner feels very annoyed at having doubted Louisiana's advice and apologizes again:

"Come on, go on, tell me everything! I certainly dug my heels in, but I'd like to continue using my intuitions wisely... especially in my profession."

“Why talk about intuitions?” interrupts the homeless man. “They only really exist after four or five of your ‘as usual’ moments, and you can tell yourself, too, that there’s another bloke like you, but smarter than you, who lives somewhere else, who’s watching you, and who gives you those famous ideas you take for intuitions.”

Surprised, Louisiana turns towards the homeless man, who continues speaking.

"We have to live in different times at the same time, whilst knowing that those other times are imperceptible and, therefore, seem not to exist, but that gives us ideas!’

Louisiana is stunned:

“See?” she says to Jacques, “even the beggars know that a double is with us.”

Jacques sighs and wants to end this boring conversation as soon as possible. He takes the bill to pay:

“The double, always the double!”

“Why assume that a poor bloke like me doesn’t reflect on his bench? He’s got nothing else to do. A time that doesn’t exist allows you to live better elsewhere. I’ve wasted too much time teaching stories from the past. Anyway, it’s better to live this bloody future before experiencing it, because that’s how we make real history.”

"That’s incredible, even the homeless know that we create the future before we live it."

Jacques gets nervous:

"Yes, but I don't live like a beggar on a bench with a future that doesn't exist."

He gets up and goes to pay at the counter.

"And how do you know that?" asks Louisiana to the beggar.

Everyone stares at this girl talking to herself. No one dares to smile, as she has come accompanied by the Chief.

"Do you know the story of the novelist Frederick Dard, the author of Inspector San Antonio? In his time, his detective novels set the mood in henhouses like yours."

Louisiana gestures for him to speak more quietly, but he raises his voice and says:

"It's a friend of a very well-known novelist who told me this true story. He'd just finished writing a story about a kidnapping. Now, his daughter was kidnapped under exactly the same circumstances described in his manuscript."

"Yes, I know," Louisiana interrupts.

She observes the surprised faces of the bar's customers.

"Best of all," the beggar continues, "was that the cops, who aren't exactly known for being the brightest, managed to find the kidnapper and the girl, unharmed, thanks to the details described in the manuscript."

Without adding water to his pastis, the beggar takes a good swig before continuing:

"Frederick Dard was so distraught he didn't know how to finish his novel."

And, satisfied, he empties his glass in one gulp.

“He hasn’t even added water to his pastis!” she says, picking up the empty glass and staring at it in astonishment.

Jacques returns just as he puts his wallet away, the phone pressed to his ear.

“Why are you ordering a pastis if you’re not going to drink it?” she says with a sigh.

Louisiana is surprised to find that the glass is full. Jacques walks away again, talking on the phone once more. Louisiana picks up the glass and looks at the beggar in puzzlement.

“You didn't drink anything!” she says, once again surprising the people around her.

“Your friend doesn’t understand a thing; he doesn’t sleep under the stars! He doesn’t know that we can fly in the sky.”

Louisiana asks him to speak more quietly, but he carries on in the same tone.

“Everyone knows that cops never use their wings, because they’re so heavy! And besides, they don’t listen to the birds singing.”

This phrase leaves Louisiana lost in thought. At that moment, the bar falls deathly silent… a silence that she thinks does not bode well.

“You see, sweetheart,” proclaims the homeless man, “a cop never dreams, never listens to the sirens’ song or the little voices that drag us to wonderful worlds. Your friend wouldn’t survive a minute on my bench.”

Louisiana shrinks into herself when, fed up, Jacques returns while putting away his phone.

“You know, homeless people talk a lot of nonsense.”

“Am I really obliged to listen to beggars?” Jacques asks Louisiana.

"But it’s thanks to a homeless man that you found the girl so quickly, because it's because of him that we met again."

But Jacques isn’t listening to her anymore, as his phone rings again. He walks away once more.

The homeless man disagrees: "It’s not because of me but thanks to me! And without wanting to seem self-interested, I think I deserved that ‘as usual’, don’t you think? But drinking it with such a daft bloke spoil everything!

Louisiana doesn’t know where to put herself, especially when Jacques returns to the table. She has never found herself in such a situation: having a drink with a homeless person in the adjoining room next to the police station, that’s just out of place!

The homeless man senses it:

"You see," he says to Louisiana, "your friend’s buying me an "as usual", but his thoughts stink. You should explain your thing to him, that bit about benevolence; it would liven things up a bit in the “henhouse”."

Thinking that everyone can hear the homeless man, Louisiana tries to explain this point of view to Jacques:

"It’s true that a homeless person smells less than your thoughts!"

The atmosphere grows increasingly tense, and suddenly Jacques bursts out laughing and exclaims aloud:

"Are there thoughts that smell nice? Due to, or thanks to the homeless, who must be allergic to soap, I'm willing to believe it."

“Better for the body to stink than for the brain to stink!” cries the homeless man furiously. “Explain to him the energy of thoughts, instead of standing there mute, like a fish in its bowl.”

Louisiana would rather be somewhere else but wants to calm things down.
“Let’s take an example.”

“Go on, we’re listening!” exclaims Jacques, beside himself. “We’re listening.”

“The example is simple: when you take the subway during rush hour, you’re squeezed in the middle of the crowd, which carries you along without you needing to put your feet on the ground.”

“That’s why I avoid taking it at those hours.”

“Yes, but if you control the benevolence of your thoughts towards others, something very curious happens: everyone moves away from you to squeeze a little further away.”

“With a homeless person,” exclaims Jacques, “you don’t need to control anything at all, their appearance is enough!”

The lights go out one by one, the bar empties to the great relief of the inspector and Louisiana, who steps outside, happy to be able to take a deep breath of fresh air.

5

Telepathy and the logic of Benevolence

But barely has she left when the inspector explodes.

"Next time, avoid comparing me to a homeless person, it causes a lot of trouble... acquaintances will ask me about my relationships."

The homeless man reacts violently.

"Better to be dirty on the outside than dirty on the inside, you poor cop!"

And he walks away, raising a finger in the air in a markedly provocative gesture.

Louisiana is stunned, as Jacques doesn't react.

"What a bunch of friends you have" he says, "and I'm not at all happy to have seen you again."

"And the subway? You haven't..."

"Did you think I was going to do what you said?'

"I just wanted to show you that those who have malevolent thoughts cannot bear the energy you emit when you monitor your benevolence."

Jacques disagrees.

"Your benevolent thoughts don't possess any energy."

"Yet they create a vacuum around you, because all malevolent people move away from you as if you were burning them."

Jacques is speechless, baffled.

"Thought is a quantifiable energy," clarifies Louisiana, "we're unaware of this because of Einstein's famous equation, since energy is always associated with mass[1] ."

"Oh no, please, not Einstein!"

"You recover your mass the instant you lose it, as this energy exchange occurs in an imperceptible time. You lose information in the future but immediately recover the fruit. This energy exchange instantly modifies your memory and goes unnoticed. Here it is no longer Einstein who intervenes, but Heisenberg."

"You can't even imagine how little it matters to me."

"It is he, however, who has demonstrated that we possess a fabulous amount of energy in an imperceptible amount of time[2]."

"So what?"

"It is, however, this imperceptible exchange that allows you to anticipate and obtain the best reflexes. Your instincts and intuitions depend on it. Meanwhile, whether sceptics like it or not, this energetic principle makes you invisible in a malevolent crowd. You then pass through a hostile crowd. No one sees you,

---

[1] . The equation $E = mC^2$ establishes an equivalence between energy E and mass m, with the ratio equal to the square of the speed of light C.

[2] . In a time approaching zero $\Delta T \rightarrow 0$, we have an almost infinite energy $\Delta E \rightarrow \infty$, since $\Delta T . \Delta E \geq \hbar$, where ħ is Planck's constant.

except those who think like you and who, for that reason, only want the best for you."

Jacques isn't listening anymore, and cuts short this conversation that bores him. His phone rings.

"Yes, it's me! Who else would it be?... Right now?... OK, I'm on my way!"

Suddenly, in a hurry, he shakes her hand and walks away looking much more rushed than he actually is."

"You really are incorrigible!" exclaims a voice behind Louisiana.

Startled, she turns round and spots the young man on the bench.

"Benevolence is a word completely unsuited to your discourse. Leave it to those who chase after imposed ideas and who gargle with empty words. Your friend Jacques thinks you've got a mouth full of sectarian dogmas without any interest."

Louisiana tries to understand the reason for the presence of this man who dares to capture her attention. Who does he think he is? The answer comes to her even before she can ask the question.

"You helped me when I lived on earth, though I know this difficult life causes us to lose our memory. I, in turn, promised to help you. That is the whole mystery surrounding us. I was part of the eleventh tribe, that of Aries."

Louisiana doesn't say a word.

"You are from the twelfth, that of Pisces! You see, my friend, you've even forgotten that… people no longer know the reason for their lives on earth. They seek an all-powerful God, without realizing that they have everything within themselves. I was like you, and you helped me. Now it's my turn to repay you."

Louisiana only half-listens. Stunned, she has just realized that this man's body has no substance whatsoever. Her hand passes right through him without a hitch.

"Stop playing like a child! You can see me because I'm in your memory. And if I come to meet you, it's not out of benevolence or altruism, but out of sheer selfishness. Each must return what the other has lent, to restore the balance. So don't talk about benevolence, it is malevolence toward those who cannot accept that sectarian word, which has enabled so many evils and embezzlements. Man distorts words to use them in lies."

Suddenly, the man looks at her without moving his lips, but Louisiana can still hear him.

"You think you hear me, but I'm only transmitting my thoughts to you telepathically… this natural mechanism is interrupted as soon as there's a grain of sand between us. Accepting my presence and my telepathic ideas depends only on you… I have no benevolence toward you, since it's a duty. One must not mistake the effect for the cause."

Seeing the unease on Louisiana's face, he tries to soften his remarks:

"My friendship towards you is merely a consequence of our telepathy. We accept each other's thoughts without any ulterior motives. It is not pure benevolence, even though it may sometimes appear to be so. So do not confuse the whole matter!

What you call malevolence attracts malevolent beings who disturb all of humanity, which in that case becomes malevolent itself. Nowadays, on Earth, no one is aware of this insidious form of parasitism, which makes you confuse demons with angels. Man has become a servant of his own demons, which survive only because of his total ignorance of the universal laws."

"I don't..."

Louisiana wants to protest, but the young man on the bench has vanished. A flood of questions comes to her mind. Could it be possible to see those who give us ideas, intuitions or premonitions?

Reflecting on the topic of telepathy, she wonders whether, after all, it was not she who caused this curious character to vanish, by protesting in her inner self. Nobody likes to hear reproaches... much less from a telepath who doesn't need to listen to hear!

***

Jacques is on the underground near the door. The car is packed. He closes his eyes and seems to be concentrating. The passengers move away from him. He timidly opens his eyes, surprised by the result.

The subway stops at the next station. He closes his eyes again to concentrate once more. The doors open, but no one enters through the door in front of which he stands. People on the platform push each other to enter through the two doors on either side.

***

When Louisiana arrives in front of her home, Jacques is waiting for her in his police car.

With an air of indifference, a neighbor smokes the cigarette he's not allowed to have at home, standing on the pavement. A policeman questioning his neighbor adds a bit of spice to his monotonous life. Moreover, nervous and impatient, Jacques gets straight to the point.

"Explain to me, what happened on the subway? I controlled my thoughts and my benevolence… well, I tried to, and… everyone moved away from me as if I were a monster! But, how crazy… I can't believe it! At the next stop, I was standing by the door, and believe it or not… everyone got on and off through the other doors, no one through mine… how incredible! I didn't know where to put myself!"

Thinking about the comments from the young man on the bench, Louisiana replies:

"Don't talk about benevolence or malevolence; it gives the impression of a cult or sectarian proselytizing, and above all because everyone gives a different meaning to the same word, and nobody understands anything anymore. You should know that telepathy between two people works when their thoughts at those moments don't oppose each other! In the subway, people moved away from you because their thoughts were the opposite of yours."

"But it's so true that we find it hard to believe… yet it works really well… I'd like to help you spread the word about this, especially if you can prove that it's a universal law."

“I simply believe that this is the application of a universal law. This scientific theory has demonstrated that time is doubled. Indeed, it is the logic of this doubling that is at the base of everything, and which the ancients called the ‘Logos’. In ancient times, the Greeks knew that an apparent logic could deceive overly trusting men. They had developed a teaching with the aim of showing that the Logos was not merely logical, but that it had to be accepted by all. It was the school of the Sophists. Everyone knows the famous sophism: “A cheap horse is rare. Everything that is rare is expensive. So, a cheap horse is expensive.” It is in seeking an incomprehensible god —and thus irrational— that we have lost this universal law.”

“When I think that I called your father crazy.”

“And you also said, “Like father, like daughter.””

But today Jacques is no longer the same. His face lights up; he has just realized something else. It is true that it would have been much easier for him to contradict Louisiana and mock her subway example. For him, the importance of thought is limitless: yes, indeed, he had created what he had thought.

“Right! I’ll think about it… See you later!”

And he leaves with a knowing wink… but immediately returns, reversing with a great noise.

“But tell me!” he adds, “when everyone knows how it works, it’s going to be a horrible mess!”

I firmly believe that this is the application of a universal law. This scientific theory has [illegible] time is [illegible]. Indeed it is the logic of this doubling that is at the base of everything and which the Greeks called the "Logos". In ancient times the Greeks knew that this apparent logic could [illegible] over [illegible]. They had developed a reasoning with the aim of showing that the Logos was [illegible] logical [illegible] it had to be accepted [illegible] the school of the [illegible]. [illegible] A [illegible] Everything that is [illegible] So [illegible] [illegible] universal [illegible] this is the universal law."

"What I think that [illegible]"

"And you," Louis said, "[illegible]"

But [illegible] the same [illegible] realized something else. It is only that [illegible] been [illegible] contradict Louisiana and mock her [illegible] own [illegible] what he had thought.

"[illegible] you [illegible]"

And he leaves with a [illegible] returns, [illegible] with a great noise.

"[illegible]," he adds, "[illegible] never [illegible] going to be a terrible [illegible]."

6

The priest's son and Distorted Truths

The next day, Louisiana passes by the bench again, where the homeless man is waiting for her with a mocking air. It's impossible for her to avoid him.

“Good morning!” she says, wondering why she hasn't changed her route, and continues on her way while waving her hand.

Passing her, a passer-by thinks she is greeting him and responds with a smile, turning around, intrigued.

“Don't you want to rest your backside on my bench today?” shouts the homeless man, irritated by her behavior.”

Louisiana sighs inwardly, stops and gives a slight grimace as a smile.

“The young lady is ignoring me, just as she ignores the fact that her lies from yesterday are the lies of those who seek to lead us to marvelous paradises.”

Louisiana stops, hesitant. She doesn't know what to do. A little further on, the passer-by tries in vain to understand what is happening.

"Sit down for a moment! That'll give you a rest. I'm sure you need it. You always have to give the lower half a break when the upper half is working too hard!"

Louisiana gladly agrees, suddenly relaxed, as the tone of his voice has become jovial.

"Are you one of those guides who are never in a hurry to go to paradise, and who prefer to send there all those who won't put up with their nonsense?"

"What's that supposed to mean? Do you have something against me?"

"I thought we were on first-name terms! But maybe you just fancy playing princesses and bloody bastards…"

"You told me I shouldn't have allowed myself to address you informally."

"I've got a right to talk rubbish too! Or do you have a monopoly on it? You know, I know all the ins and outs of hypocrisy; I'm the son of the priest's maid and also the priest's son… but you don't say that… so you know that all those stories about love, compassion, benevolence, peace and love and the like just roll off me, pure nonsense, sweetheart! Give me ten euros and I'll tell you a story of real madness, a proper trip!"

Louisiana takes out her purse.

"Did you know my father got married because of me? I found out… from my mother… about priests getting married."

He takes the five-euro note Louisiana hands him.

“With this, you’re only entitled to half the story and half my beer, unless you don’t give a damn.”

“No, tell me!...”

The homeless man holds out his half-empty beer can and adds:

“My father didn’t even know that religious marriage had been instituted by Rome. Do you know when?”

“Yes, in 1215, replies Louisiana, not knowing whether she should drink the beer to please him or hand the can back exactly as he gave it to her.”

She doesn’t notice the surprised faces of passersby who, seeing her gesticuring and talking to herself on her bench, look at her disapprovingly.

The homeless man whistles in admiration.

“Bravo, beautiful! You deserve to be on the podium because not even the priests know this. So, you can imagine how abandoned my father felt. “Do not separate what God has joined together!” he always said to newlyweds, under the pretext that Jesus had said the same thing two thousand years earlier. My poor old man, he didn’t understand a thing anymore. Do you know where that union comes from? From the double, sweetheart... God has united you with your double. The same goes for the ladies. Even the priests’ maids have a double. I told this to my parents, just imagine how stunned they were!”

Louisiana is dumbfounded.

“You know about the double?”

A couple stops, intrigued.

"What do you want?" asks the man.

"Ten euros!" replies the homeless man.

Louisiana simply smiles at them and the couple walks away, shrugging their shoulders at the same time.

"The double," continues the homeless man, "is no nonsense but the absolute truth. We must not separate from our double! The rest is pure invention by the priests. Do you know Jesus?"

And without waiting for an answer, he carries on:

"Back in his day there were some real bastards too… doctors of the law who thought they knew everything and wanted to catch him out! They were sick of this Jesus who had an answer for everything and who was seen everywhere performing so-called miracles."

"Why do you say so-called miracles?"

"Because that was a natural law for Jesus. He knew how to fix the future before living it. If you fix the future of a paralyzed man, he no longer has any reason to remain paralyzed. And Jesus could say to him with full knowledge: "Rise and walk!" Oh dear, now I've forgotten what I was going to tell you!"

He pauses for a moment before picking up the thread of his thoughts again.

"Oh, right, I remember now! I was telling you about the doctors of the law, who, to catch him out, ask him the unanswerable question: "Should the adulterous woman be stoned?" It's a trap because if he answers no, he's not following Jewish law… but if he says yes, he's ignoring Moses' law: "Thou shalt not kill!" See, the bastards… Every age has its fair share! But Jesus has an answer for everything because he's far from being a fool, especially when one is always in touch with his double. And he simply replies: "Your ass stories don't matter to us! The only important thing is not to separate what God has joined." And bang! Everyone fell silent, because in those days people still knew about the double, and they knew what he meant when he proclaimed to all those fools to shut them up: "I am in the Father and the Father is in me!" And you all are not, ha ha!"

Amused:

"I thought I was the one giving little speeches…"

"You know, a history teacher learns a great many things, truths that have been forgotten over time. There are always great fools who want to twist everything to their own advantage, to show that their foolishness is the only acceptable truth. And do you see where we find ourselves two thousand years later?"

Although surprised by such a speech, but seeing that she is running late, Louisiana gets ready to leave. The homeless man "senses" this, as he has seen her check the time.

"Don't you want to know what my father did when he found out about all this?"

"Found out about what?"

"About everything the Roman Catholic Church taught him, did you hear me? I'm a priest's son."

A passer-by startles and slows down to listen to the homeless man.

"Life isn't easy when you arrive naked in a presbytery with no doctor or midwife. A priest delivering his own son isn't trivial, is it? Can you imagine the bishop's face when he heard the news? Father René de Foulonge and his son Samson… My name is Samson, because he pulled me out of my mother's womb by the hair. Now, Samson's strength is in his hair. Samson De Foulonge, initials SDF. No Fixed Adress. You see, I was predestined!"

"Okay, keep going!"

Confused, a passer-by glares at her as he walks past.

"As my father could no longer bear all the lies they'd filled his head with, he married my mother at the town hall!..."

He stops; Louisiana senses his emotion at this memory. He quickly picks up the thread and continues:

"He even invited his bishop to the event, who, obviously, didn't come, as he suffered from a terribly sacred case of hives, totally worthy of a bishop!"

This memory still makes him laugh.

"You know that in Jesus's day, the first child in the mother's womb was considered to be the marriage. Furious at having learnt falsehoods to become a priest, my father changed jobs. He became the caretaker of the block of flats where my mother worked as a cleaner. But he still made the sign of the cross whenever I made him angry."

"Where did you learn that?"

"What? The sign of the cross?" asks the homeless man, puzzled.

"No, the double!"

"All the research I did for my old man... I studied to get out of the shit. History teacher at twenty-two. Language as polished as my shoes, suit, tie and the rest... a useless and futile achievement... now, I prefer to speak as I did in my childhood."

He falls silent again, lost in his thoughts, but soon continues:

"You see, culture's no use at all if you don't know how to sit on a bench and stop thinking. I needed it loads after what I discovered in the Gospels. Do you Judas's Gospel?"

Louisiana doesn't have time to reply as he's already launched into his speech, a mixture of passion and rage.

"That's where I found something to laugh at… it talks about the God who is in each of us… and says that it is he who helped Jesus to leave his "human garment" because they were such good friends. The story of the traitor is just a tale concocted by priests for anti-myth believers and anti-Semites. I'd like one of those myself!"

"An anti-Semite?" asks Louisiana, bewildered.

"No, just a refreshing sip of my beer before you finish it off, because you get thirsty when you're talking to a surprising nutter who actually takes the time to listen to a beggar!"

Too surprised to think, Louisiana hands him the can. The other man takes it without even realizing she hasn't had a single sip. After downing it in one go, he adds:

"And the four Gospels, do you know why there are four? My father and I looked into it. It's because of the "procession" of the equinoxes!

"I've only ever heard of the precession of the equinoxes..."

"What are you saying?" asks a young man, drawn in by the beauty of this girl sitting alone on a bench.

"Nothing," replies Louisiana, flustered.

Feeling awkward, the young man continues on his way.

"I was saying," repeats Louisiana, "that I only know about the precession of the equinoxes."

"I hope so, lovely! I just wanted to know if you were really listening to me. Don't mistake me with a ragamuffin, though culture is what remains when you've lost everything, and an old teacher still retains a modicum of "wisdom"."

And without pausing, he adds:

"In the Bible, there are six days and six nights. Did you know that?"

Louisiana nods.

"So, with days and nights of more than two thousand years, you have a week of twenty-five thousand years. Are you following me? Capricorn, night! Sagittarius, day! Scorpio, night! Libra, day! Virgo, night! And so on. And now you get to Aquarius, the seventh day of rest, man. Got it? Are your brain cells not getting too tangled up?"

Louisiana is amazed by all this knowledge. In truth, her father's theory of doubling explains this cycle and also the beginning of Aquarius.

She remembers that TV program when her father tried to explain this cycle:

"The end of the sixth night of Pisces and the beginning of the seventh day of Aquarius were triggered by the solar explosion of 13 March 1989."

And on a screen, 3D images of the doubling movement and its cycle are projected.

Just like the twenty-four-hour days and nights, the day of more than two thousand years serves mankind to create their potential future. The double is then in the future. The night of more than two thousand years, serves to fix what we have created during the day. In which case, the double is in the past. And we are in its future.

This alternation requires the arrival of what the Greeks called an "anatar". Today we talk about avatars without knowing what that represents. Between the day of Aries and the night of Pisces, Jesus arrived, the twelfth avatar, charged with bringing the doubles from the future back to the past, to maintain the possibility of incarnation on earth until the end of the cycle, that is, at the beginning of Aquarius.

Could this homeless man be informed about this law of alternation that imposes the incarnation of twelve avatars?

“It’s all your fault,” he says, “I don’t dare burp, and this beer really deserves a good burp. To finish my “procession” story, you should know that the gods come to earth for a thousand years during the ages of Scorpio, Leo, Taurus and Aquarius. And you see, when you burp you feel good and say “eu”, just like the ancient Greeks who knew how to take time to drink before talking nonsense.”

“You know Greek?” asks Louisiana in surprise; without realizing it, but clearly taken aback, she reverts to addressing him formally.

‘You haven’t forgotten, have you, that I’m a priest’s son, raised by priests amongst priests… well, those four thousand-year periods brought a message about being well. And do you know what that message was called in Greek?”

“Simply ‘ana-gelia’ or ana-gelion.”

The homeless man looks at her, nodding.

“I thought I was the only one who knew that. So you also know that ‘eu-anagelia’ or eu-anageliov is simply a Gospel, that is, a message from the gods about being well… good news!”

He stares into Luisiana’s eyes, pleased with the effect he has achieved. Her lack of reaction, however, leaves him dumbfounded.

“Do you know that we have already begun to live the fourth and final Gospel? After the six days and six nights of the doubling cycle of, it is the ‘seventh day’ of Aquarius. We are heading straight for a thousand years of well-being. A golden age, as the millenarians used to say. Are you aware that the Mayan calendar connects the last two Gospels?”

"And also, the beginning of the next cycle, but nobody knows that yet."

"You know all that too? Yet smart-alecks like you think too much and know nothing; you're just too stupid! The extraterrestrials are at our doorstep... ...and so is the good news!"

And he burps to punctuate his speech with a very loud "eu".

"The smart-aleck, too much of a bitch, knows that the four Gospels were represented by the Eagle, the Lion, the Bull and the Angel."

"Yes, but tell me, why isn't the first Gospel represented by a scorpion but by an eagle?"

"The first extraterrestrials is an exploration just as dangerous as a scorpion's sting. Now, the eagle isn't afraid of it. It lets the scorpion sting its talons, to drain the scorpion of its venom before feeding it to its young."

Not to be outdone, the homeless man interrupts.

"As for the Gospel of Aquarius, it restores our true head, that of the angel. Can you imagine the look on my father's face when I told him all this? I was discord incarnate. Every time I told him things like that, he'd shout: "You're the devil!" And I'd say: "Like father, like son." You know what he did in the end? He married my mother and died happy. Come on, get lost! I can tell you're eager to leave. Are you ashamed to be with me?"

"Not at all! It's just that I have to go."

"I want to tell you one last thing! Your business about keeping a watchful eye on the benevolence of your thoughts is really nothing more than a telepathic way of connecting with people. Those who are like you appreciate you. The others flee from you, whatever you look like, and I'm in a good position to tell you that… So, avoid talking like priests! Remember, it's the son of a Catholic priest who's telling you this… I've tried telepathy with babies passing by in their prams... and it always works. They smile at me and it's a great joy."

7

A very Erudite Homeless Historian

"You say that in ancient times, nobody was unaware that the "gods" came to visit us periodically."

A well-known journalist interviews Louisiana's father on her television program.

"That's correct! By studying the doubling cycle, I discovered that it included four physical singularities: different times crossed every 5,130 years, for 1,080 years, that is, four times during the doubling cycle of 25,920-year, corresponding to the cycle of the precession of the equinoxes."

A visualization of the movement is projected on the screen, explaining the physical mechanism of these crossings.

"For us on Earth, this crossing occurred on a very special plane of our solar system, the plane of the Earth's orbit called the ecliptic. Curiously, these singularities were called "gospels" in ancient Greece."

It seems evident that the reason for these four "good news" must have been completely forgotten by the beginning of our era. The last doctors of the law, the Platonists of the first centuries AD, knew that there were four periods of a thousand years, called the Golden Ages.

"Were they able to prove it?"

"For them, as for the Mayans who speak of four suns, the idea of the "millenarians" did not come from a teaching of no interest. When tradition required four pieces of good news, the early Christians tried to explain it using the logic of the time."

"But for them, asks the journalist, didn't the life of Jesus seem to be the only good news from heaven?"

"Indeed, and we can understand why at the end of the second century of our era, Saint Irenaeus tried to justify four accounts of the life of Jesus:

> *"The Gospels can be neither fewer nor more than four; for there are four regions of the world in which we live, and four principal winds of the earth; therefore, four are the pillars of the Church..."*

Against the Heretics (3.11.8)"

"So," concludes the journalist, "the explanation he gave in all good faith was proof of his profound ignorance of what had been taught by the "doctors of the law of time" of that era, respectful of traditions."

"And it is thanks to ignorant men like Saint Irenaeus that Mark, Matthew, Luke and John became, over the centuries, the four evangelists, characterized by the eagle, the lion, the bull and the angel:

The four zodiacal or evangelical periods.

As for the “gods”, they fell definitively into the oblivion of time, just like the other writings on the life of Jesus.”

***

The rest of the interview resolves numerous questions but inexorably prompts us to ask further questions…

“These “four gospels”, each lasting a thousand years, are periods all the more remarkable for enabling the use of a fantastic energy: anti-gravity! Hence the ability to lift enormous stones. Building gigantic monuments such as the pyramids became child’s play.

In 1998, this energy (66.6% of total energy) was revealed in the universe by two astrophysicists who, without knowing it, shared

JPGM's doubling theory. They are Saul Perlmutter and Brian Schmidt, who were awarded the Nobel Prize in 2011 for their research in this field. Indeed, they proved the existence of a repulsive force or unique anti-gravity capable of explaining the acceleration of the universe's expansion.

According to those who have long deciphered and studied the Mayan calendar, from the end of 2012 onwards, this astonishing energy could begin to manifest itself even before it is fully at our disposal. It is this energy that balances the energy of gravity during those famous four gospels. And this is also what allows the "Gods" from the sky to appear on Earth…

Those who do not accept the existence of these beings "coming from elsewhere", commonly referred to as extraterrestrials, put forward an aberrant theory but one that reassures them! It is the called the "rare Earth" hypothesis. Among all the stars that populate our universe, only ours would have a planet on which life could develop. Some even add that it is on this planet that God would have sought perfection by creating man to his image. This desolate theory still has, however, numerous adherents.

Do men not follow the same law of incarnation even though they are different?

Why would the stars not follow the same law of the time doubling, even though they are different?

Taking advantage on these four gospels, why could not their inhabitants leave their star to visit others with the simple aim of discovering other evolutions?

And everyone would enjoy this for at least a thousand years. This would finally reconcile the supporters of spontaneous creation, sometimes called "creationists", and the staunch defenders of risky evolution, so dear to Darwinists, who, consequently, do not fully understand Darwin's work.

Without evidence to the contrary, the Mayan calendar would perfectly separate two gospels of 1,080 years by a period of 5,130 years and would therefore not take into account the precession of the equinoxes. This would prove that it was made for mankind by the "gods" of the future.

On the other hand, the Hebrew calendar indicates that 5,400 years would pass between two gospels; thus, the cycle of (5,400 + 1,080) x 4 = 25,920 years of the doubling theory would be the cycle of the precession of the equinoxes, never explained but perfectly observed. It would have been established by the "gods of the past", who, for the Jews, were in the "celestial Jerusalem".

The year 2012 of the Mayan calendar would therefore correspond to the year 5772 of the Hebrew calendar. This means that the creation of "Adamou" by six goddesses, as recounted in the Sumerian tablets, would have taken place 372 years before the end of the 3rd "Gospel of Taurus", which also lasts 1,080 years.

A remarkable coincidence. This genetic "mixing" by the "gods" has just been discovered by our biologists: our DNA would therefore come from six women. It is hard for us to imagine prehistoric man carrying out such important scientific work, with a flint at the back of his cave...

The year 2012 would therefore be a key year for the Sumerians. If this period really began in December 2012, we could expect rapid and prodigious evolution, no doubt comparable to that which made the construction of the pyramids possible.

Although this may displease Jacques and those who think like him, isn't the invasion of the gods of the "fourth gospel" of Aquarius already happening? If so, anti-gravity should already be a source of evolution and technological revolution. And what about UFOs? Shouldn't they also have a place in this story?"

8

Pyramids and crop circles

The next day, a big surprise awaits Louisiana. Jacques comes round almost to apologize for his behavior the day before.

“Everyone’s entitled to their own opinion, aren’t they?”

“You know, I’m going to tell you something, as a friend, because after all, I really appreciate you… I hope you don’t doubt that. But you are a real pain! I know what you say is true, but you’re such a pain that I don’t even know how to argue with you. You’ve got an answer for everything and that’s really annoying.”

“Don’t get the wrong idea! I have answers to the questions most people still ask themselves today, and that allows me to ask myself even more important questions. But that doesn’t interest people who don’t have, or who don’t accept, the answers I’ve already found.”

Jacques is perplexed.

“Does Miss ‘Know-it-all’ really have any unanswered questions?

“Like everyone else! But they aren’t necessarily the same questions that everyone asks themselves every day… and especially those who are still searching today for a God capable of solving their problems…”

Jacques isn’t listening; he’s too preoccupied with what he has discovered.

"Last night I watched your lecture online and I must admit that I was impressed... The doubling movement, the antigravitation energy... It fascinated me! And it also got me thinking about crop circles, what you call crop circles... Do you have any idea what those things are?"

"Of course I do! If you're interested, I can show you some fascinating things, though, undoubtedly... extraterrestrial!"

"Come in if you've got a moment."

***

As soon as they enter her office, Louisiana hurries to show him a curious drawing that has nothing to do with the famous "crop circles".

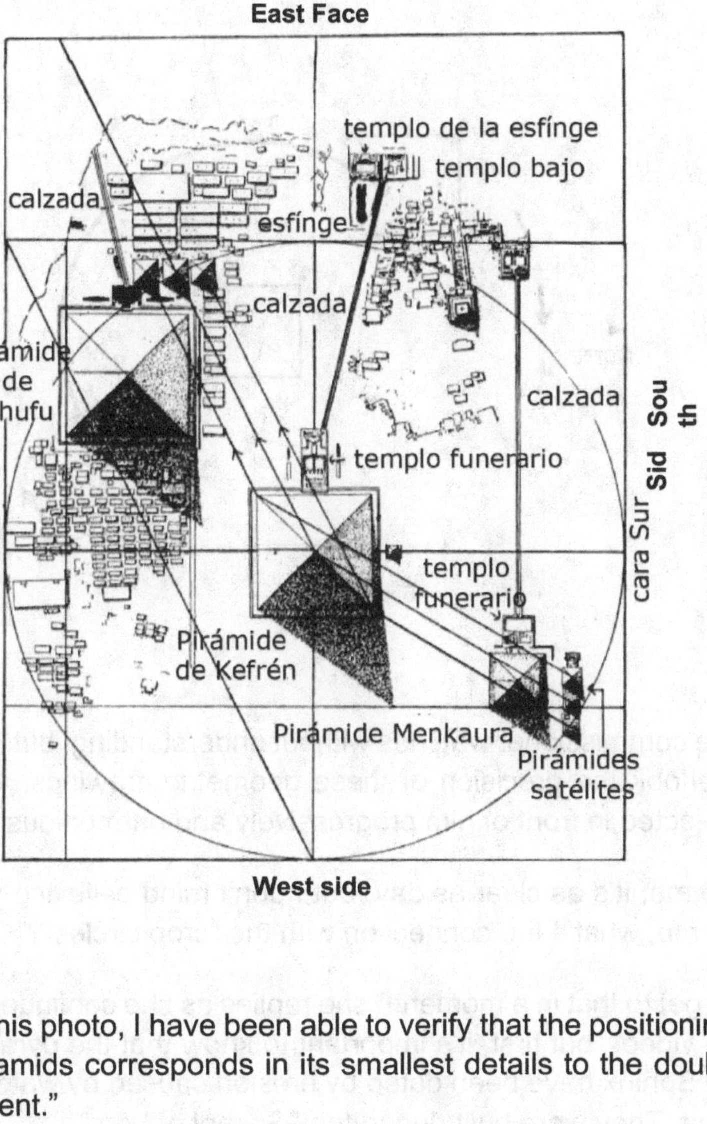

"From this photo, I have been able to verify that the positioning of the pyramids corresponds in its smallest details to the doubling movement."

Louisiana shows him a video:

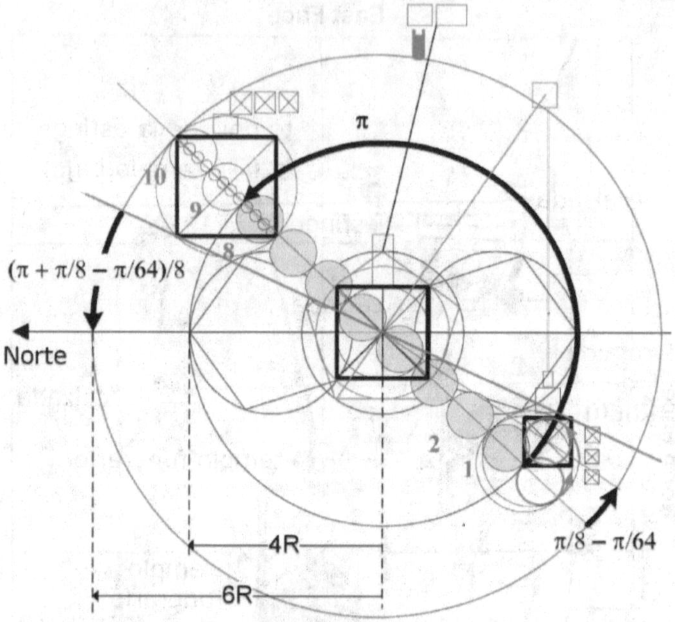

The commissioner watches without understanding, but he cannot overlook the precision of these geometric drawings, which are projected in front of him progressively and harmoniously.

"To me, it's as clear as day, but I don't mind believing you. Now tell me, what's the connection with the "crop circles"?"

"I'll get to that in a moment!" she replies as she continues to show the video, "but first, it's important to know that the pyramids and the Sphinx have been dated by erosion caused by water and not sand. They were built during the "Gospel of Leo.""

She continues with her explanations while magnificent images are projected on her screen.

"This dating was made possible thanks to the research of Zecharia Sitchin, known primarily as a novelist. Hence the criticism from Egyptologists, paleontologists and various other "-ologists"! Taking this erosion into account, the site is dated to

12,500 years ago, that is, to the zodiacal age of Leo. It is therefore normal that the Sphinx could be a lion, even though the head was added later, probably by Cheops, after the period of the "Gospel of Taurus", more than 4,500 years ago, during the IV Dynasty."

"Another French researcher, Georges Vermard, was also able to demonstrate[3] that the location of the pyramids corresponds to the constellation of Orion, and here lies the significance of this demonstration - during the zodiacal era of Leo."

"This does not mean, of course, that Egyptologists officially accept all these conclusions! They completely undermine numerous well-known publications, both one and the other! For my part, I have only noticed that in ancient Greek, Orion meant "the boundary of the gods". Is this not scientific proof that supports the exchange of information between doubled worlds?"

And without giving Jacques a moment to catch his breath, she continues:

"Evidence that undoubtedly adds weight to the mill of truth. This was also what gave me the idea to look for the doubling movement in this construction."

---

[3] . Research by Georges Vermard: We must go back to 10,500 BC for the correlation to be truly correct, a date corresponding to the moment when Orion is at the lowest part of the precessional cycle.

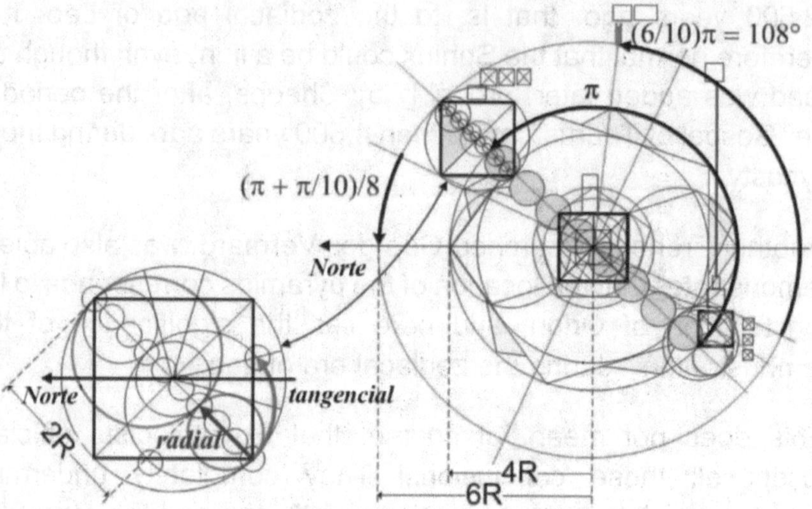

A succession of images follows, whose mathematical precision does not escape Jacques's "watchful" eye. His conservatism takes a significant blow:

And she continues with other figures that complement the previous ones.

"You realize there's a clear connection between the pyramids and this "crop circle", which to me are the very illustration of the time doubling cycle. There's nothing esoteric about it! Don't forget that around 80% of English crop circles have been made by humans."

"That's good for tourism, isn't it?" Jacques points out.

"Only about 20% of those curious countryside drawings haven't been made by humans…"

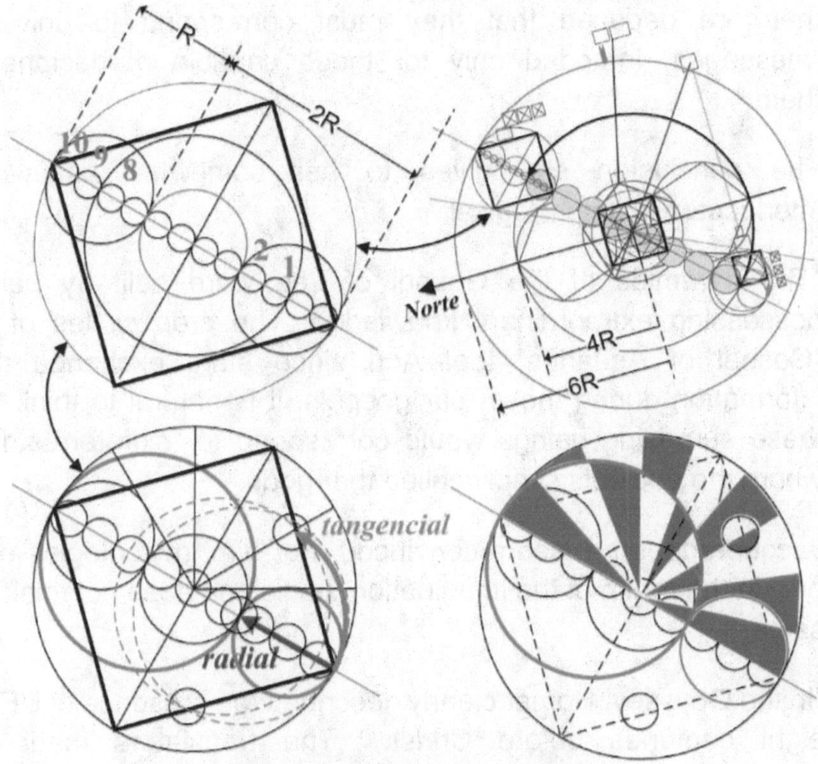

And while a multitude of crop circles scroll across the screen, Louisiana continues:

“It is not a hypothesis; it is a certainty based on indisputable scientific tests. Indeed, the real ones are all strangely related to the doubling movement, but the fake ones are not. I have therefore deduced that they must correspond to universal "messages" intended only for those capable of deciphering them."

The conclusion she gives to her somewhat astonished interlocutor is simply logical:

"The pyramids of the Gospel of Leo were built by beings possessing extraordinary knowledge. The crop circles of the "Gospel of Aquarius" too! And since stars exchange their information during these four gospels, it is natural to think that these surprising beings would correspond to "extraterrestrials" whom the ancients simply called the "gods.""

Jacques does not seem convinced. For him, mythologies exist only in the realm of the imagination. Louisiana does not think the same:

"In the Odyssey, Homer clearly described the presence of UFOs, eight centuries before Christ… The translators made up nonsensical stories that Ulysses would not have recognized. However, the translation[4] is simple, if we don't try to interpret it."

She picks up a book from her library:

"This is what the king of the Phaeacians, who came to see him with the Argonauts' ship[5], says to Ulysses, shipwrecked on his island:"

> *"'Tell me what your country and your people are, so that our ships may take you there with their thoughts (...) for we have neither helmsman nor rudder as in your ships (...) whatever we think, our ships find it on their own, they cross the seas swiftly,*

---

[4] . Translated by Ioannis Anastasiou.

[5] . *The Odyssey:* Rhapsody θ = 9, lines 555–563.

surrounded by a cloud and a mist, never fearing to sink or suffer harm."

And Louisiana concludes:

"Without a doubt, this translation strongly supports the theory that the journey of Ulysses and the Argonauts took place in a space where the stars are islands, one of which is the white light (Phaia) that heals (akesis)... hence the word Phaeacians (Phaiakes)."

Doubt once again clouds Jacques's face:

"So why don't those extraterrestrials contact us directly?"

"Obviously, those are not their intentions, otherwise, they would send us completely understandable messages. In my opinion, there are several possible explanations."

"Which ones?"

Jacques's curiosity is no longer feigned:

"Go on! This is getting exciting..."

"The "gods" come from the future or the past and are forced to wait for the equality of times to be able to manifest themselves. This equality only occurs at the end of the time doubling cycle, which the ancients called the end of times. This differentiation of times, which our science still ignores, would explain the often-surprising appearance of these UFOs."

Enthusiastic but unable to stay, the commissioner is forced to say goodbye to his friend.

Louisiana would have liked to explain to him that the Great Pyramid was also an astronomical clock, used to measure the time of the precession cycle of the equinoxes, how the diagonals would indicate the constellations of the "four gospels" and would allow us to know that we are now in the Age of Aquarius.

She would have liked, above all, to talk about the chaotic potentials of this dangerous period, for it is urgent and necessary to know how to fix them, to be able to alleviate them as quickly as possible and avoid experiencing unprecedented catastrophes.

9

Potential futures and Eucharist

After this encounter, Louisiana decides to go and see if the homeless man is still on his bench. She would like to get her wallet back...

But the bench is empty. She sits down on it for a moment. A voice startles her:

"I told you to watch the birds and listen to Jacques. They will answer your questions much better than I can."

She discovers, sitting next to her, the young man from the park bench in the flesh.

"But," he adds, "on the condition that you have at least one question waiting for an answer..."

"Actually," sighs Louisiana, "my head is full of you. You're here, and then suddenly you're not!"

"Too many answers don't allow you to get the question! Jacques was one of them... haven't you understood why yet?"

"Jacques? But you'd told me about a publisher, a springboard!"

"I guide you, because always, in the three times that are at our disposal, we are prince of one and servant of another. We mustn't forget anything when we use them as we do right now. We need it to live and survive, wherever the music of uncertainty and the wind of understanding take us. Look!"

A violent wind raises dust and makes Louisiana close her eyes and she seems to hear distant music. When the gust calms down, she finds herself sitting with the young man, but on the park bench where they first met.

"I know your concerns because they resemble my own. You think you're in one reality when, in fact, you live astride three worlds. You think yours is behind another's mirror, when we always live in front of one and behind the another!"

Louisiana doesn't understand where her interlocutor is getting at. He merely adds:

"The first advice was absurd, but it led you to Jacques. If you'd grasped it quickly, the future would have understood it too, and before you. That's why you would never have met Samson de Foulonge who, alone could lead you to the police station. Do you understand the precautions we must take when navigating the future?"

Louisiana again feels shivers running down her spine, now tingling all over her body.

"But who are you?"

"Look! Look at those birds! They know me very well. They don't need to see me to know I'm here..."

Just as happened the first time, Louisiana would like to get up but cannot. The sparrows come to peck at the grass at her feet. One of them, emboldened, inspects her shoes. Louisiana wants to bend down to stroke it. The sparrow flies off instantly, very wary, dragging the others after it.

Faced with this scene, the young man sighs, shaking his head. This only provokes Louisiana, who still doesn't understand what is happening to her. Could she be reliving the same scene all over again?

"Why am I here?"

"Listen to me! You'll understand when the time comes. Look at the birds; they know that your gestures don't match your thoughts. You frighten them! I just told you. You're only hearing the contradictory sound of your question. Like the birds, you have to listen to the answers to questions you don't have time to ask yourself. Otherwise, you'll never take flight. Living before you've truly lived is pointless, unless we decide to use it to get to the heart of the matter."

"And, in your view, what is truly essential?"

"I interpret music you are unaware of, and I am unaware of the melodies that soothe your spirit, because we do not live at the same time. But if we are at ease together, we exchange all our potentials. I know everything about you, and you know everything about me. It is the exchange of charism."

Louisiana starts thinking about the homeless man's burp, and promises herself she will check whether it is true that in Greek the "eu" of the burp means "to be well."

"You will also discover that the exchange of charisms is "eu-charist", which makes us happy in the depths of our being... A good burp after a good meal manifests the reality of this exchange! It is a good thing."

Louisiana is surprised:

"Do you mean that the tradition of burping after a meal comes from the Eucharist?"

"It's the same origin! But before burping, one must watch one's thoughts, so that they do not hurt anyone. It is therefore necessary to exchange our best potentials with all those who eat with us and who must do the same. The Eucharist is then perfect. And the telepathic exchange places everyone on the same level of intuitive knowledge, in the same well-being. Then, without intending to, the burp comes naturally! In ancient times, it was a sign of respect and good manners towards the host."

The young man pauses to watch a bird come and peck at some crumbs on his hand, which have appeared out of nowhere. Immediately, others gather at his feet.

"Do you understand that by sharing its pleasure, a single bird satisfies an entire aviary? The "Eucharist" is for them too. The abundance of exchanged potentials gives a sense of satisfaction to those whose spirit sways in a natural benevolence. Everyone is satisfied, yet only one has eaten. And, without rebellion, envy or jealousy, it is possible to survive the famine and wait for better days."

Louisiana then thinks of Jesus's last supper with his twelve apostles. To exchange potentials with that great master of thought must have been magnificent.

"Especially", says the young man, who has guessed Louisiana's reflections, "thanks to the twelve zodiacs of the twelve apostles, who opened all doors, through the harmony of their thoughts."

She then recalls the homeless man's speech about the story of the traitor Judas, which was just a story made up by priests for anti-myth and anti-Semitic people.

"Yes!" continues the young man, "without the benevolent telepathy of each individual, the "Eucharist" is not possible. And from the fifth century onwards, people began to evoke the presence of a traitor among the twelve. Now, a traitor would have "prevented" that telepathic exchange… An exchange that the imminent death of their Master made absolutely essential."

"An exchange very poorly misunderstood later on! How many errors of interpretation there have been, whether voluntary or involuntary! And how many complications in explaining those errors, trying to rediscover the original truth. And that was when dogmas replaced logic."

"You could say hello!"

Surprised, Louisiana discovers the homeless man in the young man's place.

"You're sitting there like a rude person, without even saying hello! You're babbling on about who knows what... You're the one acting like a beggar!"

"But I've been here longer than you..."

The homeless man stares at him like a pig stares at a knife.

"I don't give a damn about you! You sit there without saying a word and want me to believe you didn't see me. Take a look and see what you are sitting on!"

Louisiana stands up, spots her wallet and tries to understand.

"Are you saying I've just sat on my wallet?"

"Yes, sweetheart! As if you were afraid someone would steal it. Not a word, nothing! And you've scared all my friends away."

"Friends! What friends?"

Louisiana doesn't understand.

"The birds! They've flown off flapping their wings."

Louisiana realizes in astonishment that the argument with the young man hadn't had time to exist.

"You thought I'd flown off again..."

"Do you fly with them?"

Now it is Samson who is puzzled.

“Who are you talking about?”

“The birds! And you’re telling me you can fly too?”

Samson bursts out laughing:

“But darling, you’re out of your mind! I’m talking about the wallet Did you think I’d stolen it again?”

“Yes,” replies Louisiana without thinking. “Well, no! I mean…”

“Don’t talk rubbish! I left because of the cops, and when I came back, the young lady had disappeared and I found your wallet on the ground, under the bench. I thought you’d come to see me so I could give it back to you. But, well, I must be hallucinating! The young lady arrives, without looking at me and with a face like a thunder…”

“Yes, it’s true I thought you’d stolen my purse, just like on the first day!”

Samson’s face darkens abruptly.

“You’ve let me down, you know! You’ve really hurt my feelings... like a lash!”

But this actor adds, mischievously:

“Well, but I did take your cash, the temptation was too great... But with you, we’ll get it back in no time... Let’s use telepathy on the passers-by… okay? Shall we do it like the other day?

Louisiana is distracted. She’d like to understand what’s worrying her most at that moment.

“Tell me something: did you know the commissioner before you met me?”

“What are you on about? Do you think a commissioner would be interested in a good-for-nothing like me? We’re just scum to him. Why are you asking?”

“No, nothing! I’m trying to find a cause-and-effect relationship between you and him.”

“It’s just a waste of breath, sweetheart… without cause or effect!”

“Do you know that today is the summer solstice and that the night will be long?”

“No! It’s the shortest night.”

Samson sighs:

“You really take me for a fool! Our ancestors knew perfectly well that there were two days in the year when it was best not to sleep. Those two days corresponded to two specific positions of the earth in relation to the sun: solstice in June and in December. And, to avoid falling asleep, they would make a big fire with joy and dance around it. And when one doesn’t sleep, the night is long, man! Especially during the winter fire, that has now become the Christmas celebration... You know that, don’t you?”

“No, I only knew about the summer bonfire, on St John’s Eve. People sing and dance all night.”

“Yes, but at Christmas you have all those who are good-for-nothing but ready-for-anything types... who drag us into their supposed charity, feed us their leftovers and take us to overheated gymnasiums. That makes them feel good… they feel good inside and leave us feeling bad inside.”

The silence becomes heavy.

It is Samson who breaks it:

"Do you know why they did all that?"

"To please you, I suppose."

"I'm not talking about the jerks who decorate our Christmas night, but about our ancestors. Why shouldn't they sleep during the summer or the winter solstice?"

"Because during the solstices, our dreams drag us into very dangerous worlds that are lost in the mists of time. But who taught them this?"

They look at each other at the same time, with a sudden air of complicity:

"They knew it all thanks to the "gods" who came to earth in the last "Gospel of Taurus". This is no longer our case. Besides, everything was done to make us lose our memory."

"It is true that, to suppress these pagan festivals, the 4th century Christianity made Christmas the anniversary date of Jesus's birth..."

"But these festivals caused many nocturnal disturbances in the priests' minds. You shouldn't fall asleep, my dear! That's why they invented the three midnight masses..."

Samson continues without a trace of a smile.

"One mass alone is enormously sleep-inducing! So, with three in a row, everyone falls asleep!..."

Seeing that Louisiana doesn't appreciate his teasing, he adds:

"That's why they gave bells to the altar boys. They had to wake up the priest! Speaking of priests, I'm off to see a friend of my father's. Want to come with me?"

10.

Old Memories

Samson stands between two buildings.

"It's here!"

A small, charming church appears, completely overlooked between two towers from the fifties, a time when concrete was all too much in the minds of young architects.

"As the priest says, they're the church's two "concrete pillars": made of concrete right up to the eyebrows. They frighten him a lot, they could collapse one day, just like his priest… my father's friend."

***

Once inside the church, Louisiana is surprised to see Samson dip his hand in the holy water font and cross himself.

"Do as I do! It makes him feel good…"

And Louisiana spots the old priest sitting in the choir, with a book on his knees.

"I was expecting you!" says the priest to Samson, who embraces him.

"I've come with a "questioner" who's wracking her brain like us. But watch out! Like my dear father, she could cause a great episcopal scandal!"

A spark of hope and joy flickers in the old man's eyes.

"Do you think I can tell her…"

"Anything you like…" replies Samson. "Louisiana has already written so many things that everything else pales in comparison. I've already told her that the birth of Jesus on Christmas Day was a simple invention of the Christian and Roman Church in the fourth century, which made little difference either way."

"What good does it to you to hurt people's sensibilities?" protests Father Antonio.

"But, Father Tonio, don't classify my friend among the "easily-wounded" people. She's a reasonable girl, just like me. You can tell her everything you know, everything you've read in the distant, ancient archives of the Vatican."

"What surprised me most in the Vatican is the sheer number of books banned by the Church that have never been read. Their pages never knew the paper-knives essential in those days."

"We don't care! What matters to us is the rest."

"But do you realize that they would forbid a work without even knowing its contents?"

He takes Louisiana as a witness.

She simply clarifies:

"That fashion hasn't changed, even though books no longer have the same binding."

Samson sighs:

"We don't give a damn about bindings and illuminations!"

Quite exasperated and without waiting for her reaction, he turns to Father Tonio:

"You used to tell me that in the old days everyone knew that Jesus was the twelfth and last avatar."

"No!" he says, pleased with the effect he's created.

But, seeing Samson's annoyed expression, he quickly clarifies:

"He is the twelfth "ana-tar".

Louisiana appreciates this clarification, which she herself had also pointed out to Samson.

"The twelfth "anatar"," continues Father Tonio, "is Jesus, who comes to incarnate at the end of the two thousand years of Aries and the beginning of Pisces. The other eleven have come in each of the transition periods between two signs of the zodiac, to help our doubles pass from the past to the future or from the future to the past."

"You see," says Samson to Louisiana, "he too speaks of the double, just like the Vatican books."

"It is simply about the law of time doubling, whose movement imposes this universal rule. The role of the twelfth is to bring back to the past the doubles of men, who are necessarily in the future during the Age of Aries."

And, pleased to be understood, Father Antonio adds:

"For Platonians, it was redemption, essential to continue with incarnation on Earth, because during the Age of Pisces this was only possible with the doubles in the past."

Louisiana is delighted:

"That is where the term day and night comes from, which we find in the Bible. During the day, our thoughts create the potentials with the doubles in the future, that is what they are there for. In the night that follows, they are fixed by the doubles in the past."

The priest seems lost. He pulls a face. Seeing this, Louisiana adds:

"Be careful! The past I'm talking about is a present for the doubles who live in it, in slow motion… for our life unfolds over a much longer period of time."

The old man leads them into the sacristy, a much more suitable place for this discussion.

"In the same way," continues Louisiana, "our future unfolds in an endless time of time for those who live within it, creating our possibilities. But for us, in the time that we observe much more quickly, that potential is instantaneous."

It is then that Father Antonio, with great determination, takes a thick book with many bookmarks from a large cupboard.

"Here it is!" he says, opening a page and reading:

"A single day with your Lord is worth a thousand years of yours."

This entirely mathematical precision is found in Sura 22, verse 47 of the Quran.

He is satisfied and proudly shows his old book:

"The Quran is fine, but I hide it because it might disturb my holy water devotees… if by misfortune, they were to find it here. These days, our poor Jesus doesn't even have the right to visit his friend Muhammad!"

A leaden silence falls over the sacristy... A silence that speaks for itself.

***

It is getting late; the church must close its doors. Delighted, Louisiana invites Samson and Father Antonio to the pizzeria on the corner.

"It would make sense," says the latter, "that Jesus was born around 20–25 March, between Pisces and Aries, since he is the twelfth "ana-tar". Everyone agrees in placing him six years before himself.

"First miracle!" shouts Samson, bursting out laughing.

"All of them," exclaims the priest, "all priests' sons are hellish. Do you know why? Because they're sent by the devil! It's obvious… Just look at Samson! Thank goodness his mother was a saintly woman. That's the compensation."

He takes a bite before continuing:

"Jesus comes to descend into hell, where the infernal solutions of our future are forged."

Louisiana can't help but interrupt him:

"To do so, he must die on a specific date, during the 90 years transition between Aries and Pisces. This period allows for the birth of children who have their doubles, either in the future, like those born earlier during Aries, or in the past, like those who will be born during the Age of Pisces."

"Why," asks Samson, "do you speak of a specific date for his death?"

"During those 90 years, the future "opens" its doors for 3 years every 30 years. To succeed in his "mission", which consists of fixing the future, the best thing was to die between the ages of 30 and 33. If he died before then, the "doors" being closed, his mission would become impossible. If he died later, a third of those awaiting him in the future would no longer see him. Between the ages of 60 and 63, he could only recover two-thirds. And, between 90 and 93, one third. After that, it would be total failure.

The priest looks dazed and hides his emotion:

"That is then the reason why Jesus stayed alive until the age of 30, but in hiding so as not to risk being killed. He died at the age of 33 and benefited from a public life of three years, as tradition and some books to which I have had access tell us."

A much more precise calculation shows that he had to die at 33 years minus one-thousandth of that period, that is, almost 12 days before his 33rd birthday. Jesus came to earth to fix the "potential future" of men, which the pagan Greeks of the time called "sin[6]". At that time, all doctors of the law knew that a "messenger from the past" had to incarnate at that moment to erase the "sins of the world[7]." Indeed, that period was propitious for coming with a double in the past and a double in the future. Which made Jesus[8] say: "I am in the Father and the Father is in me."

Curious, the priest interrupts her:

"How can you advance such things, based on clever calculations? What proof do you have?"

"A completely unexpected piece of evidence! During a conference in Brussels organized by Michèle Cédric, a well-known Belgian television journalist, I met another speaker, Raouf Oufkir. Do you remember General Oufkir's failed coup against the King of Morocco? Raouf was his son."

"Yes, I remember! The general committed suicide... with three bullets in the back of the head. Well, that's what those who didn't believe the official version said. His wife and children were sent to a prison in the desert..."

---

[6] .Amartis: a thought creates a potential future, that is to say, a sin that becomes a temptation for the one who is going to actualize it in his life.

[7] . Between the periods of Aries and Pisces, the transition period is 90 years. Wide open at the beginning of this transition, the "doors" of the future close again every 30 years for 3 years. They are therefore only fully open between the ages of 30 and 33. Jesus could die between 60 and 63 years, in which case a third of the future is inaccessible. Between the ages of 90 and 93, only a third remains. After this transition period, the doors close again and the future is no longer accessible.

[8] . Jesus has a double in the past and another in the future. This is not our case. Our double is in the past so that we may incarnate on Earth during the Age of Pisces. We do not have a double in the future, hence the danger of living on Earth during this period.

And conversely, during the Age of Aries, men had their double in the future and none in the past. Jesus had to bring the doubles from the future back to the past, to maintain incarnation on Earth during the Age of Pisces. Since the solar explosion of March 13, 1989, we have been living in the transition period between Pisces and Aquarius.

"And about twenty years later, the most incredible thing happened. Locked in his cell, his spirits at rock bottom, Raouf saw a bright light in front of him. At that moment he heard a voice: "*I am Yeshuash, do not despair, you will be set free at the age I was when I died…*""

"Truly very strange!"

"Do you think that a devout Muslim, sustained by his faith during his very long captivity, could have heard the Aramaic name Yeshuash[9] in his prison cell?"

"That is also the ancient Greek pronunciation for Jesus."

"And what interested my father most was knowing Raouf's age when he was released. At the end of his lecture, he rushed over to him to ask him. His reply came straight away: "That's easy to remember! I was 11 days away from turning 33." You can imagine my excitement!"

The old man seems lost in his thoughts, and a faint smile appears on his face as if he were satisfied.

"What I like about this story, besides the confirmation of your calculations, is the fact that a Muslim feels reassured by Jesus. Isn't that a wonderful demonstration of solidarity between these two great prophets? It has been said all too often that, since Muhammad came after Jesus, he merely repeated the same good words. In your account, Jesus comes to see a man who prays to Muhammad. I love this ecumenism from beyond! How do you explain it?"

Samson gives his unambiguous view:

---

[9]. That is also the pronunciation of Jesus in Ancient Greek.

“To my father, who spoke plainly, the two were like hand and glove.”

The priest nods:

“Your father was a bit of a “rough”, but he was so genuine and sincere... A true friend!”

“For my part, I would say that Jesus and Muhammad are merely two fingers on the Creator’s hand.”

11

Science and Spirituality

This crucial topic prompts them to continue their conversation.

“For me, a Catholic priest, there is no difference between Jesus and Muhammad. But I don’t know how to explain it. It might sound silly, but that’s how it is... I really liked Samson’s father because, to rattle the complacent conservatives, he “dug” everywhere, and always ended up with a logical explanation, though often revolutionary. Now tell me what you think.”

“Oh no!” exclaims Samson, “if you ask her a question, she’ll give you an endless sermon. Be careful!”

The old man merely smiles, whilst questioning Louisiana with his eyes.

“I do, in fact, think there is a perfectly logical explanation, one that any sensible person capable of freeing themselves from dogmas, taboos, postulates or prejudices would find perfectly acceptable. In any case, we must start from the idea of a Creator who doubles himself to know himself without changing, at a time when he remains himself. His creatures will fabricate a potential future that he will accept or reject at the end of the time doubling cycle.”

Samson shakes his head in disapproval.

"As a history "teacher", accepting an initial creative force is the most scientific thing there is. Even with his Nobel Prize, Max Planck spoke of it during a lecture in Italy[10] ."

"It is obvious for those who study matter and the universe," admits Louisiana, who immediately adds: "They ignore the religious dogmas that force atheists to put on blinders. To know himself, the Creator creates creatures in his image. These find equality with Him through information exchange in imperceptible times. This generates a constant "telepathy" between all creatures connected with the Creator. There can therefore be no hierarchy among telepathic creatures, since each possesses the potential of the other. It is thus a permanent Eucharist between them! That is the reason for this exchange of potentials!"

The old man rejoices, delighted with that conclusion, but also wonders:

"Certainly, Muhammad said there was no differences between him and Jesus. But these two prophets did not have the same life at all!"

"Indeed, they came from the Creator's world in very different times. Jesus was the twelfth "avatar", arriving between Aries and Pisces. Muhammad closed the time lock that Buddha had opened for Jesus. The conditions were entirely different[11] ."

Father Antonio interrupts, perplexed:

---

10 . *'As a man who has devoted his entire life to the study of matter, I can tell you this: matter as such does not exist! All matter arises originates and exists only by virtue of a force that causes the particles of an atom to vibrate and holds this tiny solar system of the atom together... We must assume that behind this force lies the existence of a conscious and intelligent Mind. This Mind is the Matrix of all matter."* Max Planck, lecture in Florence, Italy, 1944.

11 . See the book *Change Your Future Through Time Openings*.

"Are the different epochs the cause of the difference in possibilities?"

"Of course! Muhammad's arrival on earth follows the doubling law[12]. He too was able to see the future in order to fix it before living it. That was his role as a prophet, for in his time, the future remained very violent. Knowing the law of the ages, Jesus knew that the first thousand years of Pisces would be difficult."

Samson, without blinking:

"Was it already like that, a thousand years before?"

"That's what we think! But it is easier to live during the day of Aries with the double in the future and guardian angels in the past, than to live during the night of Pisces, with the double in the past and no protection in the future."

"But why?" asks Father Antonio.

"Guardian angels do not exist in our future."

"No theology mentions this."

The homeless man sighs. The historian wakes up again:

"But you know perfectly well that no current theology speaks of the double... Thomas, apostle of Jesus, is nevertheless very clear in his Gospel, which has evidently not been recognized by the Church[13] ."

---

[12] . Jesus was born in 6 BC. The Aries-Pisces time lock opens 540 years later (that is, 5 times 90 years) and closes again 90 years later in 624. Muhammad died in 638, 12 years after the lock closed. The opening of the lock occurred 540 years before the birth of Jesus. We place the date of Buddha's birth between 624 BC and 563 BC. The analogy between the teachings of Buddha, Jesus and Muhammad clearly shows that these three prophets used the same source of information.

[13] . Found in the Ethiopian desert in the 1940s. *'On the day when you were one, you became two (for 24,840 years). But when you become two, what will you do?'... "When you make the two into one, and make the inside like the outside, the above like the*

"And the Templars knew they were not alone on their horses, but that they were with their doubles."

The old man looks thoughtful.

"According to what you just told me, after the night of Pisces we find ourselves in the day of Aquarius, meaning we rediscover the guardian angels in the past and our doubles go to the future."

"No! It is simply the end of the time cycle. Whatever the cost, we must reunite with our double, wherever it is. This time, Jesus must help us to pass into the past of our double, which has remained in its world, to go to one of the dwellings he has prepared for us[14] . We then leave our earthly world."

"And does a scientist dare to speak like that?"

"This is what my father teaches, and it is what leads some "cold-hearted" scientists to make foolish confusions between science and spirituality. Science and the cycle of time are at our service, not the other way round. Why not push physicists who comfort themselves in a narrow-mindedness that makes them feel at

---

below,... when you make male and female into a single one, (for 1,080 years), so that the male will not be male and the female not be female...then you shall enter the Kingdom." Gospel of Thomas (sayings 11 and 22).

[14] . *"In my Father's house there are many rooms [...] I am going to prepare a place for you. And after I have gone and prepared a place for you, I will come back to take you with me, so that you also may be where I am."* Gospel according to Saint John, 14:2–3

ease when faced with such a marvelous universe? There is a creator of the doubling. It is the law. Anyone who ignores it outside the law and, therefore, becomes dangerous without knowing it."

"Isn't this the case," the priest asks with a slight raise of his eyebrows, "of all those who refuse to recognize the existence of that Creator?"

The homeless man sighs again.

"A Creator of our doubling has nothing to do with the God of religions, which drag us towards dogmas of no interest whatsoever. Jesus has told us: "There are some standing here who will not taste death." It is evident that all of Jesus' contemporaries are dead. But this phrase has a very different meaning for the doctors of the law. At the end of the time cycle, he will come to find us to save us "the weeping and gnashing of teeth". And that makes me feel good when I'm dying of boredom and cold on my bench."

12

The Death of the Beggar

The next day, Louisiana is tidying up the attic in her building. She plans to offer it to the homeless man. She hangs one of her father's suits on a nail and leaves pajamas, a shirt and socks on the bed.

She takes teddy bears, dolls and other things out of a box… and sits on the bed, lost in thought.

As she leaves the building with two large bin bags, Louisiana spots a police car in the distance. The neighbor, always the same one, with an air of indifference, is smoking the cigarette he's forbidden to smoke at home on the pavement. This apparent indifference turns to watchful attention when he spots Louisiana. There is no one nosier than this man, who knows all the neighborhood gossip and feeds it day in, day out. He's having a right good time of it. A policeman making Louisiana get into his car is a clear sign of misconduct! He's already thinking about all the things he'll be able to tell…

He doesn't realize that we can live out whatever we imagine. He thinks his mediocrity is the result of bad luck. It would never occur to him, on the other hand that he might reap what he sows with his horrible gossip.

***

Louisiana isn't feeling very well. She has just been taken away in a police car like a common criminal, under the gaze of her neighbor. The two officers — the same ones again — don't dare raise their voices. They don't understand anything either, but they don't want to face reproach from their chief commissioner. As soon as they arrives, he receives her in his office and apologizes:

"I didn't mean to cause you any trouble by sending a police car to fetch you like a taxi… It's about your beggar. He's dead…"

"Dead! But how can that be?"

"The question isn't 'how?', but "who?'"

Louisiana doesn't understand this question, which seems completely out of place to her.

"You should know he was a history teacher at the Henri IV secondary school. A good man! He died alongside his wife and the mother of his son in a car accident."

"Dead?"

"Dead and buried by his son."

He never told me about his son, but I knew his name: Samson de Foulonge."

"His name was Samuel, not Samson!"

"How curious! But that doesn't change his initials. He was proud of them."

"His initials? Oh yes" Jacques realizes, "SDF, No Fixed Adress (Sans Domicile Fixe), but a history teacher! This makes no sense… let's check all this out… it's far too strange."

***

The next day, by chance, Louisiana finds Samson on his bench. What a surprise!

"I've been told you're dead!"

"There are always bastards going round telling people what they think they know. You don't understand a thing about our shitty freedom, so stop giving me a hard time! My spirit is my freedom. A body is a prison. You know that! Have you seen all those bastards who think the world will end on 21 December 2012? What will they think when they're in 2013 and they're still alive? The end of a cycle isn't the end of the world."

"It's the spirit of truth that will invade everyone."

"Spirit of truth! Then, few will resist, can you imagine? Everyone will live the consequences of their lies, their beliefs, their hypocrisy."

"We're going to have a good laugh!"

"And why not change our way of thinking right from this moment? We could then experience all the good that our imagination would create."

"I like girls like you!" says the homeless man in a mocking tone. "They dream! Nothing else will happen… at most, the human "stupidity" will be even more explosive!"

"Why waste time on some poor bloke who doesn't understand a thing! Imagination is the only way forward, the only way to get us out of the mess our past thoughts have got us into. So imagine the best instead of going round and round in circles over the same old nonsense! The gates of the stars are opening, and only those who have no beliefs, no prejudices, but who simply wish to live a better future will enter. They must, however, have been able to imagine it, even if they're dressed in rags that reek of wretched poverty!"

She stands up, disappointed.

"By the way, it seems Samuel de Foulonge is dead and buried... So, who are you, Samson?"

The homeless man also stands, sighing.

"What a silly question! I had changed Samson to Samuel to look more serious at school. But I do have a fixed address where you can check that I'm still alive and kicking. Do you hear me? The entrance is on Boulevard de Ménilmontant, opposite Rue de la Roquette. You can't miss it. It's the third street on the left in that housing estate built for people like me. My son has taken me in again. No more homeless man under the starlight!"

And he walks away without looking back.

Louisiana does the same, but realizes she's forgotten to ask him the number of the boulevard, and goes back in vain in the beggar's footsteps. The street is deserted. How could he have vanished so quickly?

A phone call to Jacques and she gets the address... It's Père Lachaise. So, she decides to turn left at the third street. Suddenly, she stops and stares in amazement. No, she isn't reading it backwards:

"Here lie Samuel and Aimée de Foulonge."

In a state of shock, she sits down on the grave and seems to hear the homeless man's laughter:

"I have a fixed address where I'm very much alive!"

She is surprised to find herself crying, though without any sadness, as if a great joy were overwhelming her. A young man approaches her, very intrigued.

“Excuse me, do you know my wife?”

“No, only your father… And I didn’t know he’d died…” How did it happen?”

“An accident… and If I had had money, I’d never have buried him with my mother and my wife.”

Louisiana then discovers the name Sybille de Foulonge beneath the other two names.

“You should know that I have no desire to meet his friends.”

Louisiana wonders if they are talking about the same de Foulonge. She doesn’t want to leave without making sure.

“Your father was the history professor at Lycée Henri IV, wasn’t he?”

“Yes, that’s right! And he was very good at telling stories... Could you leave me alone, please? It’s the anniversary of my wife’s death and I’d like to be alone...”

Louisiana takes a fifty-euro note out of her wallet.

“Please put some flowers on it from me, as a remembrance.”

As she hands the fifty euros to the son, Louisiana recalls her first meeting with the father. He suddenly seems so close to her that tears of emotion roll down her cheeks.

Seeing this, the son can’t help but utter a perfunctory remark:

“Why get so worked up? He wasn’t worth anything like that.”

Reading genuine compassion in his face, Louisiana bursts out laughing... And she thinks she hears, like an echo, the homeless man laughter.

***

As she leaves the cemetery, Louisiana comes face to face with the young man from the park bench, who is waiting for her calmly. About twenty pigeons circle around him. Stepping aside, passers-by surround him, giving him an air of reality that he lacks, except in Louisiana's eyes.

"Do you understand now why your publisher is no longer interested in your manuscript? Your book will be written with Samson in mind, and your conclusion will be completely different from what you thought you had written."

"And what is that?"

"More or less well-informed, the "enlightened ones" of all kinds talked about the end of the world in December 2012. They fell silent in 2013, just as the numerous false prophets who predicted enormous catastrophes for the year 2000 have remained silent."

"We are, however, living in very dangerous times, aren't we?"

"Alarmists predict potentially dangerous futures created by human thoughts. That is their role. Prophets of good or bad omens must show you the need to change your life. To do this, there is no need whatsoever to philosophize or discourse about the gender of angels. Let's imagine a wonderful future! Yes, we always live in the fable that we tell ourselves and dream about, even before we live it. Soon everyone will do the same and, like most dreamers, they won't know why or how, and there will be vain despair. It is now up to you to invent everything necessary to dream well!"

The pigeons fly away and Louisiana realizes that the young man is no longer there.

13

A Very Much Alive Dead Man

Intrigued and turning over in her mind the latest events she had just witnessed, Louisiana decides to return to the park to see the homeless man. Seeing him on his bench, she surprises herself by repeating what she had already said the day before:

“I've been told you're dead!”

And she hears the expected response:

“There are always bastards who say what they think they know. Have you come to apologize? Well, get a move on! I'm only here so you can look at yourself in the mirror without being horrified by your own reflection. You wanted to offer me a room under your roof, is that it?”

Louisiana is stunned. And what Samson adds with a chuckle plunges her into an even deeper state of confusion.

“Just imagine how ridiculous I'd look sitting at my desk in your father's suits. Can you picture it? I've guessed it all… So get out of here before you make me sick!”

“First, you're going to listen to me. I found the grave where you're buried alongside your wife, Aimée de Foulonge.”

“You're mad!” shouts the homeless man, completely disturbed.”

"Until proven otherwise, the living aren't buried. And your son would have preferred not to be forced to bury you alongside his mother and his wife!"

"My son!" shouts Samson, astonished.

Louisiana notices that the homeless man's mocking smile has vanished. He's as pale… as death!"

"I lost my ID when my wife died. And my son rejected me because I killed his mother and his wife. I don't even know what he's up to anymore… and what's worse, I don't care!"

And he says, growing agitated:

"Because he buried me, he thinks I'm dead! The fool, it reassures him, and you too… you wander through my dreams, that's all! Can you imagine being in the cemetery with the son who comes to see his mother and his wife? You wander through the secret gardens of a poor bloke like me or like him. If you keep sitting on my bench, you'll end up raving as much as I do."

"Why did you tell me you had a permanent address with an entrance on Boulevard de Ménilmontant opposite Rue de la Roquette?"

The homeless man looks at Louisiana with the air of a beaten dog.

"Because it's true! My son has rehoused me! Now he can live without scruples."

Seeing that Louisiana is trying, in vain, to understand, Samson explains:

"One evening I was driving my car with my beloved wife beside me, and my son and his wife in the back…"

He thinks back to that day; a great sadness is reflected in him, his gaze into the void…

"I crashed violently into a lamppost. When I regained consciousness, I saw our two women lying side by side on the pavement. A pedestrian also died, his face completely crushed. They'd covered him with my old leather jacket, which contained all my documentation. Thinking it was me, my son was cursing him, calling him every name in the book. I tried to talk to him but he wouldn't listen to me, he couldn't see me, he was so overwhelmed by grief."

He sees himself again talking to his son, but the boy ignores him completely.

"So, I decided to give it all up… I can't remember what I did… probably the shock. And one day when I was feeling down, I followed my son as he went to the cemetery, and then I discovered my name on my wife's grave and his wife's. Imagine my surprise! My son had buried me… that's why I have my fixed address in the cemetery. I knew it would astonish you, but well… I like the idea."

"What idea?"

"The idea of being dead."

"And your son?"

"He's not like you. He can't hear or see me. Not so his son… he tells me I'm not dressed properly and that I smell just as bad as his father's rubbish."

"And I was just about to suggest…"

"Yes, I know! The father's clothes and the little girl's teddy bears…"

Louisiana is stunned. How could he have guessed her intentions? Samuel smiles at her:

"Do you think you're the only one who travels in dream time? I've seen all your dolls lined up in single file on the bed you've made for me. But I think I'm going to accept your degrading proposal. You never know. If we make up, I might even get to see my wife again."

"I think first of all you should have a good bath to wash off that stinking layer of grime you're carrying!"

"All right!"

He simply vanishes, and Louisiana finally spots the onlookers, who stare at him in astonishment as he talks to thin air.

Entering her flat, Louisiana discovers disturbing barefoot footprints on the carpet.

In the bathroom, Samuel is enjoying himself in the bath whilst surveying the countless beauty products, carefully arranged on a shelf.

A ring at the doorbell makes him look up.

Louisiana opens the front door and is astonished to find Jacques standing there.

"Your homeless man isn't a ghost at all. He's an impostor. The history teacher from the Lycée Henri IV is well and truly dead and buried. We're investigating your homeless man, who probably has something to hide. Be careful!"

A noise coming from the bathroom surprises him.

“Aren’t you alone?”

“Yes, of course I am.”

A splash and an equally loud curse surprise both Louisiana and Jacques.

Jacques pulls out his gun and heads for the bathroom. He opens the door, looking menacing.

“Don’t tell me it’s the commissioner coming back to bother us again!” Louisiana hears and she gives a slight smile.

However, to her great surprise, she sees that Jacques doesn’t seem at all surprised. She opens the door and sees for himself that there is no one in the bathroom.

‘What was that noise?’ he asks

“I don’t know, maybe it was my ghost.”

Jacques shrugs, but Louisiana notices his puzzled look.

At the police station, nobody knows the homeless man.

The two policemen are adamant: Louisiana was always alone and told them nonsense about a homeless man and a bench that doesn’t exist!

Stunned, she heads to the location: the bench isn’t there anymore! The shop opposite is modern and bears no resemblance to the old one.

In the room adjacent the police station, the waiter is categorical. She was alone with the commissioner. Jacques in turn confirms everything the waiter has said.

Louisiana is going crazy. Trying to make sense of it all, Jacques discovers, to his surprise, in an old land registry, the existence of a church that has now disappeared, and obtains the address.

The both go to the location. The church between the two towers has disappeared, but they address is correct; the two towers are still there. An elderly couple who run a shop remember the church and its priest very well.

All the church's books have ended up in the municipal library, including a Quran with annotations by the priest.

***

When she returns home, she hears a noise in the bathroom again and, stunned, discovers Samson in the bath.

'You again?"

"Didn't you tell me I must wash myself to go and see my son?"

"But who are you?"

"Why won't you believe I'm my son's father? Well, I'm going to prove you wrong, so you don't go making up stories as silly as they are stupid, just like that bastard."

14

Reunion and explanations

Handsome and smartly dressed, Samson gets out of the car that Louisiana has just parked in front of a shabby looking building with rather attractive architecture.

“Are you sure he still lives here?”

“Of course!”

Samson doesn’t want to explain why he’s so certain. He has often come here to spy on his son. He knows he has a son. After the accident, before she died, his son’s wife gave birth to a boy.

“I think we should have warned them. Seeing you might give them a severe shock.”

“We’ll see!” replies Samson, who doesn’t want to show that he’s nervous, like a child in new shoes. Deep down, he implores his wife to help him be well received by his son.

Luckily, the building’s door opens just as they arrive at the intercom. Holding the door open, very politely, Samson lets a person coming out pass, and they thank him for his kindness.

Louisiana is surprised:

“She saw you”

"Above all, she opened the door… We've passed the first barrier without any trouble."

"I've told you before, we always live what we imagine. So, we'd better imagine the best!"

Samson looks at the names on the intercom. Louisiana senses he's on the alert.

"Well, we just need to open up," he says, pressing the buzzer.

No one answers. Everything is silent.

"Let me try!"

She presses the buzzer and a man's voice answers:

"Who is it?"

"I'm a friend of Samson de Foulonge. I'd like to talk to you about him. Do you have a couple of minutes?"

"It's on the third floor, to the left of the elevator."

***

As he steps out of the elevator, seized by a sudden panic, Samson runs down the stairs four at a time. Louisiana doesn't have time to react as a door opens in front of her and a tall, handsome young man appears. He is deeply unsettled upon recognizing her.

"You again?"

“Well, yes! I’ve…”

“What do you want from me?”

Looking frantic, the son’s eyes widen excessively as he glances behind her. Like a block of ice, Louisiana turns around. Samson is standing there, a faint smile playing on his deathly pale face.

“Simon, my son!” Simon turns pale:

“Dad!” he cries before fainting.

***

Samson enters the flat. Seeming to recognize the place, he opens the living room door and helps Louisiana lay her son on the sofa.

A scream startles them. Little Maxu’s nanny has just walked in. Seeing Simon unconscious, she screams before fainting herself.

“Good heavens! This is a disaster!”

“No, for me it’s beyond the grave!”

As they sit the woman down on the sofa next to Simón, a tiny voice startles them.

“Well, I never!” murmurs the three or four-year-old boy. t’s not bedtime for parents and nannies!

“And you, what’s your name?” asks Samson, very surprised but infinitely happy to be talking to him.

"Maxu."

"And you, aren't you going to ask me who I am?"

"But I already know you're my daddy's daddy. I asked for you to come and see us."

Disconcerted, Samson doesn't understand, no more than Simon, who has just regained consciousness:

"Who did you ask?"

"Myself!" he says, laughing.

"Yourself?" asks Simon, completely bewildered…

"But you don't understand anything, daddy! he says, amused, "I asked my "essential self" to tell you. That's why your daddy was able to come out of my dreams."

"But nobody asked me anything!" exclaims Samuel.

The child sighs loudly, turning to Louisiana:

"You see, even being dead, a grandfather is also a daddy, that's why he doesn't understand anything. But you, who are big like mummy, how were you able to see him?"

Now it's Louisiana who feels out of place:

"Who are you talking about?"

The child looks behind her, laughing:

"You tell her!"

Louisiana turns round and sees the young man from the park bench smiling just before he disappears.

"Why did he say that?"

"He didn't say anything," replies Louisiana, still in shock from that quick vision.

"But are you deaf or what? He said you're able to see what grown-ups can't see… because you know how to play with times and because you know it's very, very difficult to play with that. Hey, will you teach me to play with times? I love jugglers on TV… but… what are times?"

Speechless, Samson fixes his wide eyes on Louisiana. Simon does too.

"But tell me! shouts Samuel to his grandson, "I'm very much alive, aren't I?"

"Yes," replies Maxu with a smile, "but in dreams, because here you're just a big ghost. Look!"

The child walks right through him again and again, laughing.

"Grandma has been waiting a long time for you; she's waiting for you. She says you're a playful one…"

"And a stubborn one," adds Grandma, who has just appeared.

She looks lovingly at Samson who, without trying to understand, takes her in his arms, crying with joy.

His mother gestures to Simon, who comes over to hug them both. Father and son look at each other, as if they were discovering one another or finding each other again.

The nanny, who has just awakened wonders why Simon is standing with his arms outstretched, hugging the air.

"Put your shoes on!" shouts the nanny at Maxu, fixing her hair in front of the mirror. "I've wasted enough time already."

Simon startles and realizes he is alone.

"You see," Maxu exclaims to Louisiana while putting on his shoes, "nannies always think they have to save time, but then they don't know what to do with it. You, who know the time that others don't see, should stay with daddy. He loves wasting time with me. Are you divorced too? Or are you alone and sad like my daddy's daddy, who's smaller than Daddy, even though he's the grandfather."

"But you talk so much" replies Simon, amused.

"Yes, but I know what I'm saying."

Turning to Louisiana, he adds:

"I'd like to know if you're divorced."

Simon looks amused:

"Little children don't ask such questions."

“Yes, but I’m a big boy who asks whatever questions he likes.”

And without waiting for an answer, he grabs the bag his father hands him and blows a kiss into the air.

“Next week,” he says to his father, “we’ll go with her. She’ll teach you to see what you don’t want to see and she’ll make a pretty little sister for me.”

The nanny shrugs.

“What nonsense are you lot making up again?”

She looks at Simon with an amused, questioning expression.

“But nanny, it’s not nonsense,” she says, turning to Louisiana, “nannies don’t understand anything. And you, won’t you understand anything either when you’re the mummy of my little sister?”

Feeling very uncomfortable, Louisiana hurries to say goodbye to these people who have left her quite bewildered.

***

As soon as he leaves the building, the man from the park bench walks alongside her.

“Why are you running away? I’m doing everything I can to ensure your life isn’t wasted, and you’re foolishly turning your back on everything I’ve managed to prepare for you.”

"But why are you following me everywhere? Are you trying to drive me crazy?"

"If you can see me, it's because you accept my help… so, why are you getting nervous? I'm preparing the best possible future for you. Have you forgotten that you must bring Floria here with Simon's potential?"

Louisiana tries to understand, but in vain.

"She has to be your daughter so that you can benefit from my help and I from yours."

"I don't understand a thing. What are you on about?"

"If you met the father on a bench, it was so that the son could meet you and get you pregnant… have you lost your memory to such an extent? We must fix what we have disturbed! And you no longer understand a thing…"

Maxu arrives in his stroller with his father.

"It's normal not to understand anything when you're a grown-up," he says to the young man.

And like a grown-up, he explains to Louisiana:

"Before leaving with his sweetheart, my grandfather told me that you need to open the future and ask your father for the key. Then, once you've sorted everything out properly with your "Psykaios", you'll close it again."

"My what?" asks Louisiana.

Without answering, Maxu turns to his father.

“And you, why do you never talk about the Psykaios? Why?” he asks his father from the height of his few years.

“About what?”

“The Psykaios.”

“I left my Psykaios on my star to come here with you, but you don’t even know it exists.”

The three arrive at the park where Louisiana had met the young man.

“There are many things I don’t know, my little Maxu,” replies the father, “and a Psykaios on a star, I don’t know where you’ve got all that from.”

“Not on a star, but on my star in the sky!”

His father’s lack of understanding makes him sigh.

“Why is there a star of yours in the sky?”

“Like everyone else! I’ve left there to help my Psychaios fix the things we disturbed.”

“What have you disturbed?” asks Simon, smiling amusedly.

That smile exasperates Maxu.

“We disturbed the world to come. We need to put everything back in its place. That’s why we’re here…”

"You haven't forgotten your Psykaios, have you? He must be feeling terribly sad."

With that, Maxu rushes over to the bench to sit down. Immediately, sparrows come and perch at his feet.

"But darling, who's put those stories into your head?"

The child looks at his father with concern.

"I can't believe it!"

Maxu feels chilled. How could his father have forgotten the most important part of himself?

"And her," he says, pointing to Louisiana, "you don't even know that she's here to help you, just like me. Oh dear, this is serious, a daddy who doesn't remember anything in his little head..." and he adds with a very grown-up air, "it's going to be hard for her!"

Seeing that Louisiana is smiling as she looks at her father, he gets nervous:

"You shouldn't laugh at everything, or I'll get very, very angry, and when I get very, very angry, Akoké Anogui goes away!"

Simon smiles at Louisiana before explaining the meaning of that name:

"Akoké Anoni or Anogui is his stuffed animal that he often talks to."

Maxu sighs, looking at Louisiana:

"Daddies are like birds; they don't understand a thing!"

Louisiana picks up her phone.

"Excuse me a moment; my father must be worried as I was supposed to visit him."

"Why are you wasting your time on a mobile when you could be talking to the birds?"

Louisiana freezes. It's exactly the same thing the young man said to her when they first met.

Just as it happened the first time, sparrows come to peck at the grass around them. Summoning up his courage, one of them approaches, inspecting Luisiana's shoes as she bends down, smiling, ready to tame it. Suddenly, it takes flight, dragging the others behind it.

Maxu sighs, shaking his head. A sparrow approaches Louisiana and then turns back as if suddenly afraid.

"Birds know everything and are afraid. You have to pet them, but only when they ask for it!"

Amused, Louisiana thinks of the young man who used to say the same thing to her. This intrigues her deeply.

Maxu protests, displeased:

"What I'm saying isn't funny."

And as if guessing the direction of Luisiana's thoughts, he sighs:

"And Akoké Anogui isn't my stuffed animal; your daddy will explain it to you. The birds have already warned him; you'll see."

Louisiana's phone rings.

Maxu smiles:

"See?"

It is indeed her father calling.

15

The Psychaios and Akokhé Anoni

Louisiana introduces Simon and his son to her father.

Passionate about this subject, he searches in the Greek dictionary for the explanation he lacks.

At the far end of the living room, Maxu is fast asleep on the armchair, nestled between two cushions and his teddy bear.

“Have you found anything?” asks Louisiana to her father, whom she sees smiling.

“Perhaps! In very ancient Greek, “Psykaios” exists and means, more or less, “that which comes from the soul”. This could perfectly well refer to the double. How surprising!”

“Would my son speak ancient Greek?” wonders Simon, who quickly adds, “but you should know, I don’t believe in reincarnation at all.”

“Neither do we,” replies Louisiana.

‘What is re… “re-incarnation”?” asks Maxu’s little voice, who has just opened one little eye, concerned…

“It’s difficult to explain to you, little man,” replies Simon.

“It’s, above all, completely ridiculous!” exclaims Luisiana’s father. “If you fell into a stinking sewer, full of ugly rats eager to eat you, and manage to get out of there, would you want to go back in again? Well, you see, there are people who think they live several times on Earth."

"But we can't just leave our psykaios like that,!" cries the child, amused by the idea.

"Leave what?" asks the father.

"My Psychaios! And on top of that, mon would have to come back too and mom's mom."

And looking at the dictionary, he asks:

"Can you find where it is in your book?"

"What do you want me to look for, son?"

"My star in the sky. It's in Kuknos."

"Kuknos! That means swan in Greek. Are you talking about a star in the swan constellation?"

"Yes, it's a big bird that's going to bed, because it's very tired, just like me."

He lies back down on the armchair.

JL is very surprised to hear a child talking about a constellation in which a black hole was discovered at its center at the end of the last century, just before that of our galaxy.

Curious and hoping for a possible explanation, her father writes an email to a friend, asking for help to clarify this topic.

On the other hand, he refers to Greek; it could be that there was an omega between the 2 consonants and an eta at the end.

As a child, María thought she was constantly with someone like her, a twin or a soulmate, as this being seemed to be imprisoned in a body without gender or androgynous. When she walked down the street, she would move people aside to let them pass. At mealtimes she demanded an empty chair beside her. No one understood her. How great, then, was her relief and joy when, years later, she heard a French scientist speak of a double as a vital principle of life.

The reply is not long in coming:

"This child reminds me of my childhood. I think agoghe is a Greek word derived from the word ΑΓΩΓΗ (alpha, gamma, omega, gamma, eta) and means: to lead, to act. Anoni reminds me of the Greek word ΑΝΩΓΙ (alpha, nu, omega, gamma, iota), which means "higher than the earth". It is a compound name: ANW (alpha, nu, omega) and GAIA (gamma, alpha, iota, alpha)... What the child is trying to tell us is that, through his double, he receives teachings from another world, higher and more evolved than the earth... I feel a great emotion..."

"How can a child express himself in Greek?" wonders Simon.

"Why imagine silent doubles? Wouldn't they speak to one another? And if they do, why not express themselves in Greek?

More than four centuries before Christ, Plato has Kratylos say: *"I think, Socrates, that absolute truth is nothing but a force, superior to human force, that put the first names to each thing with the condition that they be rigorously exact.""*

Since Greek is a scientific language that explains the doubling[15], it is difficult to think that men have been voluntarily plunged into ignorance by suppressing this knowledge.

Often, time and foolishness are enough to make us forget what is most important. However, with knowledge of the cycle of m time doubling, the Greek alphabet takes on its full meaning and sense. It also proves to us that in the first centuries after Jesus Christ, our doubling was still a truth taught by the "scientists" of that time, who were called "doctors of the law" or "Platonians". Why were they massacred under the Vatican in the sixth century of the Christian era? Might the answer lie in the question itself?

---

[15] . By the same authors, see *Change Your Future Through Time Openings.* Editorial Reconocerse (Publisher).

Did you think this was going to end here?

You open a book, read it, close it, and that's it… well, no! We're passing the baton to you. You must now pass it on to others, but without chasing after them. Sometimes it's hard to wait in vain for someone who might ask you for help. But proselytism is a tool of darkness that should never be used.

So don't lend this book to your friends! It's a bad habit to force people to read what we've enjoyed. Let's leave them free to discover their future possibilities through simple telepathy. They'll know whether they need it or not, and whether you'll be of use to them or not. And stop helping those who haven't asked you for anything!

It's amazing how some people feel the need to build themselves up, thinking they have the answers to other people's problems, when they don't even have them for themselves. Fortunately, you've read these pages and you're no longer one of those!

The time doubling theory and the calculation of the speed of light

Jean Pierre Garnier Malet

Currently, matter, space and time are defined by the Standard Model. The Higgs boson was supposed to complete this model, but rather than sorting everything out, its discovery shows that, faced with numerous inconsistencies, fundamental physics must find a new path.

Now, **the doubling theory** explains several fundamental points, proposing and justifying a new paradigm. Based on the doubling of space and time, it implies the doubling of the observers of these spaces and times (read more).

This theory introduces a new concept of the particle horizon, in privileged observation planes (the plane of the universe, the galactic plane, the plane of the ecliptic, the plane of polarization of cells in doubling, etc.). Thanks to this new paradigm, a horizon in a plane is always a particle in its own horizon. And the particle of a horizon is always a horizon of particles. Particles and horizons follow the same doubling movement but at different scales of time and space which requires the differentiation in the observers' perception.

Three horizon dilations ($x2^3$) and an acceleration of movement from 1 to 1,000, in an observation plane of a three-dimensional space, transform 1% energy into 1000% for an initial observer, but in an imperceptible time, called a "time opening" and defined by the doubling movement. The latter only perceives the beginning and end of the doubling cycle.

These dilations and accelerations require three energies that the doubling movement defines **rigorously**: an anti-gravitational energy (66.6%), a gravitational energy (33.3%) and an equilibrium energy (0.1%). Used in imperceptible "time openings", this 1% energy becomes 1000% in another space where that time becomes perceptible to other observers evolving in that space. This is the objective of doubling.

We thus rediscover the cosmological constant of 66.6% that Einstein could not impose, forced to confess that: "It was the biggest mistake of my life!". In 1998, whilst observing a supernova, **Brian Schmidt** and **Saul Perlmutter** demonstrated this 66.6% anti-gravitational energy that accelerates the expansion of the universe, thereby corroborating the doubling theory. They received the Nobel Prize in 2013.

Doubled from this initial observer, a 2nd observer uses this energy in a time that he truly perceives, thereby carrying out an experiment in a time imperceptible to the initial observer. This real experience is a potential future, which the first observer instantly memorizes in imperceptible "time openings". We finally understand the utility of **Heisenberg's uncertainty principle** ($\Delta E.\Delta T \geq \hbar$), which ensures an almost infinite potential energy in an almost null time.

This theory imposes 3 fundamental constants: the speed of light C_0 and 2 superluminal speeds C_1 and C_2 which are linked by the equation:

$$C_2 = 7\ C_1 = (7^3/12)10^5\ C_0$$

Currently, physics imposes the postulate of an instantaneous exchange of state between doubled particles, considering that there exists a principle of non-locality.

Calculated by the doubling theory, **these superluminal speeds eliminate this postulate.** They were first demonstrated in 1982 by A. Aspect, although it was not until the experiments conducted by N. Gisin and A. Suarez that they were accepted (these three scientists were awarded the Nobel Prize in 2011). As these speeds are very large but finite, entanglement is not instantaneous.

The doubling of a particle takes place in the privileged plane of an observation horizon between an internal (radial) particle that penetrates a horizon and an external (tangential) particle that goes around the horizon, itself being an external particle in its own horizon…

This new notion of doubling (radial and tangential paths) is perfectly observable in the planetary movements of the solar system, of which we are the observers in our time of observation. In 2005, this notion led me to **justify the arrival of planetoids** in the Kuiper Belt that borders our planetary system, and to **associate spaces two** by two: Sun-Pluto; Mercury-Neptune; Venus-Uranus; Earth-Saturn; Mars-Jupiter; Asteroid Belt-Kuiper Belt.

This association continues into the infinitely large. It implies that our observable universe, as seen from within, is linked to an external space, observable only in an imperceptibly short time (**Planck time**). In other words: we do not have time to observe that external space. It can only be observed through a shift in the perception of space and time defined by the doubling theory. With this change of observation, our universe becomes imperceptible, but it is not a **black hole from another universe.**

A black hole has been defined as being a singularity of our universe (**Penrose** and **Hawking**), because it differentiates the observation time inside and outside a horizon, in a privileged observation plane. However, it ensures energy exchanges with the outside (Hawking radiation). It is only this dynamic aspect that has a reality, because the black hole is just a place of energy exchange. This is probably what **Stephen Hawking** is trying to tell us today (January 2014), surprising the scientific community once again. He thus adds water to the mill of the doubling of space and time, around a horizon of particles in a privileged observation plane, defined by the doubling theory.

This new notion of the internal radial and external tangential allows us to calculate the speed of light in the solar system, where we are the observers of that space in a time of observation.

Above all, it helps us understand the fact that this speed is independent of the observer's speed and that of the source (**paradox imposed by Einstein**).

Indeed, **it is not speed but the relationship between two** observable **times**, necessary for doubling.

For us, as observers in our observable universe, time is observable through the Earth's movement or by its time of movement. This relationship can be expressed by the ratio of distances travelled in the same time, or by the ratio of travel times.

If we want to express it as a speed in the equations of mechanics, we must choose the relationship between a length and a time. Since the doubling theory defines the internal (radial) and external (tangential) paths of our planet around the Sun, we can use this as a means of calculation with the Sun's radius Rs^{+} or of that of the Earth. In this way, any planetary space in our solar system allows us to verify this definition.

An example for the Earth, of which we are the observers:

The Earth's radial path is the distance between aphelion and perihelion

The tangential path is the rotation around the Sun

The ratio of the two distances therefore gives this universal constant: R_s^{+}

$$C_0 = \frac{\textbf{radial path}}{\textbf{tangential path one year}} = \frac{108 \times 10^4 \times (4\pi R_s^{+})}{\quad} = = \mathbf{299{,}792\ km/s}$$

This calculation uses the dimensions and travel times of the Sun and Earth (whose radius is linked to Rs +).

However, the speed of light is a universal doubling constant. If the diameter of the Sun or Earth were to change, the dimensions and travel time of the Sun, the Earth—and other planets—would be modified to keep this constant fixed. It remains linked to the universal doubling movement, whose duration of 25,920 years corresponds to the precession cycle of the equinoxes, never before explained. The doubling theory allows us to calculate it, explain it, and, above all, predict the current end, especially the arrival of the planetoids (see publications).

Conclusion

Physics must consider the existence of imperceptible time openings between horizons evolving in different times. These openings link observers who perceive time differently.

The doubling of time theory implies the existence of doubled universes, embedded in the same cyclical doubling movement. Imperceptible cyclical exchanges of paths (internal and external) provide instantaneous information without modifying the appearance of the movement. **This implies entanglement, non-locality and super-luminous speeds.**

This doubling allows us to explain: the Big Bang and what precedes it, the apparent matter-antimatter asymmetry, dark matter, dark energy, antigravity and the acceleration of the universe's expansion…

This mechanism of imperceptible exchanges of states or information in imperceptible time openings finally allows us to understand the anticipation necessary for all evolution.

The movement implies a doubling cycle (corresponding to the cycle called precession of the equinoxes, observed in our solar system), the significance of which we have not yet become aware of the importance. Divided into twelve periods, this cycle of time doubling of is coming to an end. We are living in a transitional period of 1,080 years between two cycles, and we could avoid serious planetary disturbances by understanding the fundamental mechanism of the doubling of time, instead of ignoring it.

JEAN-PIERRE GARNIER MALET

Physicist, fluid mechanics Paris - France

By the same authors

Change your future through time openings

A simple, practical, scientific book of unexpected and considerable scope. This is the book that aims to transform our sense of life.

In a simple way, in this work, we can rediscover the purpose of our life and restore order, both in our body and in the world in which we evolve.

A book destined to revolutionize the lives of its readers.

The Double... How does it work?

It is thanks to our double that we can become clairvoyant and "change our future" through imperceptible "time openings".

Reading this book, you will discover the simplicity with which it is possible to find a daily balance and, above all, the purpose of our life.